THE
Ancestral Continuum

Unlock the Secrets of Who You Really Are

NATALIA O' SULLIVAN & NICOLA GRAYDON

**SIMON &
SCHUSTER**

London · New York · Sydney · Toronto · New Delhi

A CBS COMPANY

First published in Great Britain by Simon & Schuster UK Ltd, 2013
A CBS COMPANY

1 3 5 7 9 10 8 6 4 2

Simon & Schuster UK Ltd
1st Floor
222 Gray's Inn Road
London WC1X 8HB

www.simonandschuster.co.uk

Simon & Schuster Australia, Sydney
Simon & Schuster India, Delhi

A CIP catalogue record for this book is available
from the British Library.

ISBN 978-1-84983-756-9

Typeset in Times by M Rules
Printed in Italy by L.E.G.O Spa

For Luis Pando Rivero & Brian Evans Graydon
&
To all our ancestors, near and far, named and unnamed, the
inspiration behind our lives, we dedicate this book to you.

Contents

Prologue

When I was young I used to have a recurring dream. I was wearing a blindfold and walking towards a door. When I strained to look beneath the blindfold I could see men's feet, shuffling, on the other side of the door. They were wearing dark, old-fashioned shoes; some were smartly polished, others scruffy. I remember feeling frustrated that their shoes were the only feature I could identify. I could also see a light shining under the door, which I knew I had to walk through. I would then wake up.

I never told anyone about this dream. Children don't try to understand dreams; they only wish them to stay or go away. This one eventually went away.

In my late teens the dream returned and this time I discussed it with my mother. I'd had the 'blindfold dream' again, I told her, revealing for the first time its origins in my childhood. She astonished me by confessing that she used to have the same dream as a young girl.

When I was pregnant with my first child I visited my relatives in Spain. As an expectant mother I realised I was no longer going to be a passive recipient of the gifts – and burdens – of my parents and their parents and their parents' parents before them. I wanted to know more about where I came from. My relatives took me to the grave of my grandfather and it was there that I was first made aware of his story – and its connection to my dream.

On the eve of the Spanish Civil War in 1936, my grandfather, Luis Pando Rivero, was a judge in the Galician port town of Villagarcia de Arousa, where he lived with my grandmother and their three daughters. He was also the President of the Republican political party the Frente Nacional, who were in direct opposition to Franco's Nationalist military oppression. The intellectuals of Galicia were particular irritants to the right-wing factions of the country and everyone knew that my grandfather had Republican sensibilities. As a judge he had to order the detention of many fanatical Fascists who had been accused of brutally murdering local civilians. So when Franco marched to power on 17 July 1936, his allies warned my grandfather to leave Spain immediately with his family. A large number of prominent Galicians had already fled to Argentina, but he refused to go, believing his steady voice and compassionate justice would help sanity prevail. I think he probably thought he wasn't in any danger, as he had done nothing wrong, and that would be sufficient to ensure his safety and that of his family.

However, local Fascist party members knew he had no sympathy for their actions or their political party, which is why they wanted him out of the way. They cleverly waited for their chance and then, as soon as an opportunity arose, they arrested him.

In response to a decree controlling the possession of firearms, he signed an agreement to safeguard the guns of local Republicans, saying he would be taking responsibility for them. He was trying to avoid confrontation by demonstrating that things in Villagarcia were under control, but he had inadvertently signed his own death warrant. One night in August the guards arrived at his house and arrested him. He was accused of fomenting rebellion. They used the signed agreement as evidence.

I have since found dozens of letters that he sent my grandmother from prison, written on toilet paper from a dank cell in the infamous La Isla de San Simón jail off the coast of Vigo. The early letters were requests for clothing, toothpaste or fruit; later there were robust instructions to my grandmother as to who she should seek out as counsel in his defence. Eventually the letters took on a strained, melancholy tone; even his handwriting shifted from a forward slant to a backwards slope. Though he did not express his fears to his wife, it was as if he realised that his cause was lost.

Four months after his arrest, he was tried by a kangaroo court in Pontevedra and found guilty of crimes against the State. Then, early one morning when the sun was rising above the prison walls, he was led from his cell down a long corridor, through a door that opened onto a courtyard, and taken away to a high open space above the town. There he was blindfolded with seven other men and shot by firing squad, on 4 December 1936. He was forty-four years old. As he died he heard some of the men cry out 'Viva la República!' He died in silence.

Just as I was finishing this book, my cousin Luis showed me the prison where my grandfather had spent his last hours, and the place where he had been executed. It was all exactly as it had been in my dream.

My childhood dream came back to me with added poignancy but it raised more questions than answers: what had he been feeling as he took that last walk to his place of execution? He must have thought of his wife and children. Did he die fearing what would become of them? Would he have feared death itself? Might he have been cursing himself for not leaving the country? And what did his life have to do with me? His story had some negative repercussions on his family. My cousins told me that my grandmother had a nervous breakdown soon after his death and while she was recovering their daughters were sent to be educated by my grandfather's sisters, who were nuns at an exclusive boarding school. My grandmother never spoke about what had happened; she kept it a secret, like many families in Spain. My mother also does not talk about it, but she remembers that she did not recognise her mother when they were finally reunited. So, while the light of my grandfather's heroism shines through history on his descendants, his children undoubtedly suffered from his decision to remain in Spain.

For myself and my children, I believe that his story has given me the courage to stand up for truth and justice and I have an innate belief that everything in life is possible. Luis was an extraordinary man known for his fairness and philanthropic nature. His example encourages me to continue my work in a compassionate way and helps me to understand the effects of trauma on families around the world: that is his legacy to me. I have told his story and celebrated his life with my children and with that I impart my own understanding of what his sacrifice meant for all of us.

My dream was the beginning of a journey of discovery that has brought many gifts. I cherish his memory and he has made me proud of my Spanish heritage. It has inspired me to help others discover their own extraordinary ancestors who, in one way or another, tried to make a better life for their descendants. Through him I have come to understand the value in remembering them: when we honour their journey we immeasurably enhance our own lives and those of our descendants.

This book is the result of half a lifetime's journey towards that understanding.

Introduction

The choices that we make in life are not unique to us.
They are a distillation of all that has come before us.
The more we become aware of our ancestral lineage,
the more freedom we will have to honour what is best
and let go of the rest.

DENISE LINN

We are all part of an ancestral continuum that began on the First Day. We are linked to history and pre-history by an unbroken chain of living, breathing people who fought for their survival and struggled against oppression, who won wars and lost faith, found power or wealth only to see it disappear, who died loved and unloved, who created industries and who sought ways to make a better world. We are a product of all that – the good and the bad.

Our ancestors remain alive in our genes and their memories reverberate in our imagination. If we really wish to discover who we are and why we are here, we need to remember them. If we remember them, what secrets might they reveal to us? By learning about their troubles we can begin to find ways to alleviate our own. And, if we let them, they can help us to step into our true power and fulfill our highest potential.

Long before there were computers to research family histories, our ancestors were kept alive in the stories and myths of the small villages and towns where they had lived for generations. Those who had departed were kept close: their graves were visited regularly, their achievements celebrated and their names invoked at family gatherings. Today, our lives are more fragmented and family members often move away from their place of birth and from each other. But while we may have forgotten our ancestors they remain with us, whether we are conscious of it or not.

Through dozens of interviews and personal investigations, we have discovered how the experiences of our ancestors play a major role in the way we live our lives today: they affect the choices we make in our careers, our partners, our finances and in bringing up our families. When we reach back through history and see our ancestors' journeys through time, we can start to discover our own physical, psychological and emotional heritage. And through examining what is known of our family history we can also connect with our ancestors in a way that makes them a real and vital presence in our daily lives.

For some people, it is the birth of a child or death of a parent that precipitates a desire to discover more about their origins. For others, the onset of a congenital illness or a history of depression, addiction or other emotional issues may be the starting point. Perhaps we are blocked in our careers or find ourselves unable to have children. People who no longer live in the country of their family's origin often feel compelled to seek out their roots – Americans of Irish, Scottish and other immigrant bloodlines tend to be particularly well informed about where they came from;

many have well-documented family trees and have travelled to the birthplaces of their forebears.

When we find out who our ancestors were – exploring their names, the places where they were born and lived, their occupations, marriages, illnesses and deaths – they start to come alive again. By resurrecting their memory, we bring them towards us into the present time. And as we do this work of excavating our family's past, we will also find ways to heal ourselves, our relatives and our family tree, thereby offering a legacy for the many generations that follow us.

It may be that everyone in your family has lived happy, peaceful lives. However, the past two centuries have seen more upheaval on a global scale than any other time in history, and during your journey of research you may find that some of your ancestors experienced pain and personal tragedy. Healing the physical, emotional and psychic wounds in the family's lineage might be the most important thing we ever do. We first need to discover what those wounds are and, with that awareness, begin to explore how using the ideas, exercises and true-life stories presented in this book can help. It is a revealing journey of self-discovery that will ultimately liberate us from the burden of our ancestral history and reconnect us with the more positive aspects of our own heritage.

I believe we all have a gallery of ancestral heroes and villains and in knowing them we come to know ourselves better. My grandfather is my ancestral hero. My dream was only an introduction to his life and legacy, but the more I have learnt about him, the more I have benefitted from his courage and determination to stand on the side of truth and justice. I have felt his presence with me at times when my own resolve was weak. His belief in education, law and service is reflected in all my cousins. Learning about him – first through my dream, then through the recollections of my relatives and my own research, then through meditations, rituals and other practices which we will discuss later – has made his strength and wisdom a part of my everyday life. I see him as a benevolent ancestor watching over my family and guiding us in spirit. I see that as clearly now as if he were sitting beside me while I write these words. Coming to that clarity was a journey. And that is the path we hope to share with the readers of this book.

My own life has taken a dramatically different course from that of my grandfather. In my teens I discovered I had the gift of highly developed intuition. I believe I have inherited this from my family. My grandmother and her friends held séances in my Spanish grandmother's house; she also used to do card readings to foretell opportunities for romance, money, pregnancy and marriage. She learnt these skills from her mother, who also used to read tarot cards.

These influences had a more formative effect on me than the world of law to which my grandfather devoted his life. As I grew up, my fascination with the spirit world developed with my psychic skills. Over the years, and through many hundreds of readings with clients, I have been privileged to 'hear' the voices of their ancestors. As I hear their voices I feel their deep desire to help their descendants. Sometimes I hear a guardian ancestor who shines a light of wisdom and compassion. They are often more present in moments of crisis and celebration, from the birth of a child to the death of a parent. They all come full of their best intentions and advice. For many people, this connection with their past brings peace and purpose. Sensing that they are being looked after by an invisible presence, they find they are a part of the ancestral continuum that links them with the past and the future.

Nicola, my co-author, was awakened to her connection with her ancestors through the death of her father when she was only twelve years old. It was an early initiation into the impermanence of life and it turned her into a spiritual seeker with a strong belief in life after death and the spirit world.

Born in Africa, she inherited her father's love of the wildness of the African bush, high veldt thunderstorms and the natural world. Her mother's family was also connected to the land, as her English grandfather, a self-made man, had acquired vast parcels of fields and woodland in the British countryside. This is where Nicola grew up, when her family moved back to Britain shortly before her brother was born. She was on a self-destructive path when we met in our late twenties, confused in relationships and trying to manage the polarities of the two legacies she had inherited: braais in the bush and dressy dinners at the Savoy. The connecting thread of these disparate worlds had always been her relationship with the land's most ancient inhabitants: the trees. As a girl, she found a hollow oak in a nearby park where she would go to hide when family squabbles erupted. That tree was her sanctuary.

Shortly before we met, she had ended another relationship and was in deep despair. Driving home one night, she was overcome with a feeling of wanting to die. She pulled her car to the side of the road beside a great old tree where a memorial had been created with teddy bears, candles and cards. There were messages in childish writing saying 'Too bad the party had to end' and 'The light has gone out of the world. I miss you.' She stared at it all for a long time before seeing the name of the young woman whose life had ended in that spot: it was Nicola. In that shock of awakening, she felt the spirit of her father urging her to embrace the life force that her ancestors – on both sides – had bestowed. They would be her protectors until it was her time to join them.

As Nicola discovered her relationship with her ancestors, she began her lifelong fascination with indigenous cultures and their relationship with the ancestors. She learnt about their understanding of the rites of passage from birth to death, reaching back to our most ancient times, and she learnt that their codes and practices have been passed down for generations and generations.

Every culture has a folklore tradition and belief in ways of connecting with the spiritual aspect of its ancestry. In Celtic traditions the Druids understood that there was a progressive connection between the three known worlds of nature – the underworld, the earth and the upper world, or paradise. In Africa many peoples believed the ancestors resided in the earth, before the missionaries told them to look to heaven for inspiration. They would say that the personal power we seek lies beneath our feet. The Aboriginals connect to the Dreamtime, the ancient times when their forebears sang the earth into existence. This is comparable to Carl Jung's 'collective unconscious'. It is where our most potent ancestral memories reside, the place whence our inspiration comes and the source of our dreams and visions.

This book is a journey through the labyrinth of our personal ancestral heritage, guiding us towards the truth of who we really are, and it is for everyone. You don't need spiritual beliefs to draw on the power of your ancestors. Whether you put your trust in God and an afterlife or in science and genetics, the power of those who went before you can be a guiding force in the life ahead of you. What follows is a framework for connecting with that power.

ONE
The Tree of Life

All over the world people have become disconnected from their family tree. They have been uprooted for so long that they have forgotten what it means to be connected to their ancestors. And when we are disconnected from our family tree, we are disconnected from the family tree of humanity.

MANDAZA AUGUSTINE KANDEMWA, TRADITIONAL HEALER

When we are born we arrive bearing the genetic imprint of our fathers and mothers, who carry the genetic imprint of their fathers and their mothers, who carry the imprint of their forebears and so on. The development of DNA technology has linked us once again with our distant past. The secrets of our inheritance – our gifts and talents, our predilections and habits, as well as our physical characteristics – lie in our genes. They contain the memory of all that we are and all who have gone before us.

We are also beginning to realise how much our lives can be influenced by the latent ancestral memory of generations past. Negative ancestral history can be an impediment to our personal growth, blocking our dreams and preventing us from becoming who we really are. Meanwhile, other positive ancestors – and we all have them – are in the shadows, waiting to be of assistance should we acknowledge their achievements and connect to them.

In the western world – unlike in African, Asian and tribal communities – we have forgotten how to maintain a spiritual link with our deceased family and the vital importance of honouring and celebrating our ancestors. As we remember our ancestors we once again reclaim who we are, rediscover our creative gifts and our spirituality. We become aware that we are not alone, that we are part of a continuing ancestral narrative that can lead us to our life purpose and immeasurably enrich our relationships with ourselves, our families and the wider community.

Lands of the ancestors

> Scientists now calculate that all living humans are related to a single woman who lived roughly 150,000 years ago in Africa, a 'mitochondrial Eve'. She was not the only woman alive at the time but if geneticists are right, all of humanity is linked to Eve through an unbroken chain of mothers ... all the variously shaped and shaded people of the Earth trace their ancestry to African hunter-gatherers.
>
> **JAMES SHREVE**

Human beings are migratory and exploratory in nature. We began as small groups of hunter-gatherers constantly moving with the seasons to greener pastures. Our ancestors migrated from Africa about 150,000 years ago to spread through Europe, Asia and Australia. We reached the Americas about 20,000 years ago and inhabited the Pacific Islands just 2,000 years ago.

So although we claim to belong to a country, our heritage as humankind is one of continuous evolution through migration. Our country of birth may influence how we feel and see ourselves today but our ancient ancestors lived in many different lands. In our subconscious we remember these places and, should we ever go back, they trigger some long-forgotten memory in us. It feels as though we have been there before. Africa especially seems to exert a magical pull, giving rise to the phrase *mal d'Afrique*, meaning that once you have been there you long to go back.

Since having children I have felt pulled to take them to both Spain and Hungary to connect with each side of my family tree. From the first time when I stood by my grandfather's grave during my pregnancy with Sequoia, my eldest child, it has felt important to me to honour the lands of their ancestors.

When we went to Hungary, it was my middle child, Ossian, who felt the most connected. We had gone to spread my father's ashes in his homeland. Ossian thrived in the luxuriant countryside of his grandfather's parents. Even though my father only visited rarely after moving to Britain, his family welcomed us with great love and hospitality. Family is family and when it is linked to the land of origins it feels particularly powerful. Our Hungarian family still lives on the farm that at least five ancestral generations have inherited. Like the generations before him, my cousin and his family work the land assiduously, and so will their children. The wisdom and security in knowing the place where they live are part of who they are. They have a deep sense of connection to the land: this is something my husband and I (like so many of us) have lost as we have moved around so much. I am sure that this lack of established connection is ingrained in refugees and immigrants the world over.

Meanwhile, Nicola has always felt deeply connected to Africa, not simply because she was born there but also because her grandmother's family – Huguenot refugees – have lived there for generations. Despite being brought up in England, it is to South Africa she goes when she needs to recharge. Touching the earth as she arrives, she can almost feel all the cells in her body breathe a sigh of relief.

In Britain today there are still families who have lived in the same village or area for generations and local graveyards record their names through history. In one astonishing discovery, the DNA of living villagers in Cheddar were found to be a close match to the so-called Cheddar Man, Britain's oldest skeleton and about 9,000 years old. Having roots this deep creates an innate sense of belonging. But we can all tap into that when we return to the land of our ancestral origins. You may live in Detroit or Boston but perhaps your family emigrated from Ireland to the USA. The land of our origins, our roots, echoes in our bones. The soft, green hills of Ireland remain inside you.

Sara's story

Sara Connell is an author, speaker and coach and lives and works in Chicago. When she became a parent she looked back into her family history to name their son. 'As a child, I was proud that my father, William Casey, was one hundred per cent Irish, as in America it was rare to be one hundred per cent anything. But I know only a little of our Irish ancestors in the generations before my great-grandparents boarded ocean liners for America: my great-grandmother Brigit was from Limerick in County Kerry, my great-grandfather was from Cork.

'When our first child was born we named him Finnean, after the mythic poet-warrior. This was the name in my heart during pregnancy and I wondered if it was possible for a child to whisper his own name to us. His name, whether self-chosen or given, is Celtic, and when he is older we will take him to Ireland, offering him a visceral place to explore the meaning of his name and Irish ancestry. We can go to St Brigid's Well at Liscannor and the fairy rings of Kerry, the cliffs of Moher. And, in the south, where our ancestors are from, I will walk with him into the hills covered in wet green moss and we can listen together, with our inner ears, to whatever ancestral voices we may hear in the land.'

In the year after her father Johnny Cash died, Rosanne Cash felt compelled to make a trip to Scotland with her daughter Carrie. Her father had traced their family roots back to Malcolm IV of Scotland, and to the small town of Strathmiglo in Fife, where they walked around looking for streets named after their ancestors.

The links between Cash and Scotland were musical as well. 'Going further back into our Celtic past made my father realise that this was where he derived his tone of voice, the mournful quality to his music and it was that sense of place and time that was passed on to him and then to me.' She described how inspiring it was to be in Scotland: 'It was really thrilling. The area is spectacularly beautiful. If geography can be in your cells and in your deepest memory, then it came out of the realms of the unconscious and was fully alive for me and Carrie that day.'

The trip was also immensely healing for another reason. Rosanne was browsing in an antique shop when the owner, hearing her American accent, asked if she was searching for her roots. She told him her name and he mentioned that Johnny Cash had traced his family to the town. On hearing that she was actually his daughter, the man disappeared upstairs and came back with a photograph of himself with her father. Rosanne burst into tears. 'When I realised he had been to this exact place it just hit me. It was the best gift that man could have given me. I was returning to the place where our family's story started, somewhere that gave my father so much pleasure and pride.'

Knowing where we come from is a precious gift that can influence our life journey in hidden but powerful ways. Pauline Tangiora, a Maori elder who is half Scottish, took her grandson to the birthplace of her mother in Dunbar. He brought back with him a roof tile from her mother's first home. It is, she says, his most prized possession and now he sees himself as the holder of the Scottish heritage for the rest of his family.

The pull of the homeland remains a powerful one and an ancestral pilgrimage can help us connect to our lineage and feel a sense of wholeness and completion. Terry went to Ireland not long after the death of his father, Jeremiah Patrick (known as Paddy). Paddy was born in London's East End and had always promised himself he would go back and visit his family's homeland in Ireland, but he never found the time. After he died, Terry felt an overwhelming urge to go to County Cork. 'It felt like a pilgrimage for him. I took his photograph with some small mementoes to lay in that soft earth. I dug a hole with a trowel, just deep enough to insert my memories of him and the lives we shared together. I ushered the earth across, burying the images reminding me of his youth, the times he spent as a seaman, a firefighter, father and grandfather. I lit a candle, left Irish whiskey in a container and prayed to the ancestors' clan of O'Sullivan to receive him home. I felt the wind blow off the Atlantic and a seagull called as if it were the last post.'

Going back, he felt his ancestral homeland in his bones and he has been back several times since. Last time he went to County Kerry and found himself visiting a graveyard near his hotel. 'It is filled with memorial stones to all past parishioners, many buried with headstones announcing this region "was strongly inherited by the clan O'Sullivan". In Cork I felt as though I belonged but when I went to Kerry I felt as though I had come home. In Kerry, surrounded by plot after plot of O'Sullivan graves I felt elated, bright, animated. It felt like a real homecoming.'

Indigenous traditions say that the ancestors reside in the earth and as, through the ages, we have mostly buried our dead, this is literally true. In a spiritual sense you feel them most strongly in the land of their origins. This is where it is easiest to connect with them. Certainly, as I stood beside my maternal grandfather's grave, I understood him – and myself – in a visceral way; equally, when I went to Hungary, I understood my father better and saw where his great warmth and loving energy had come from. As I spread his ashes, I grieved for him once more as he had been unable to return to his homeland himself.

Migration

In the last three centuries we have seen waves of voluntary and involuntary migration, as our ancestors leapt continents, fleeing persecution or seeking economic prosperity. In these days of air travel, phones and the internet, it is difficult to imagine how epic those journeys were. Emigrants left their homelands with few belongings for an unknown future, not knowing if they would ever see their families again.

Some 12 million people entered the USA via Ellis Island in just thirty-two years. In some cases entire villages packed up and booked a passage to New York City as stories spread of the opportunities to be had in the New World. At its peak in 1907, over 1 million immigrants passed thorough the halls of Ellis Island. For many, their name was changed and women, often arriving years after their husbands, were given new clothes and a new hairstyle to reflect American fashions, so their husbands would not be embarrassed by their old-world style.

When Nicola's husband Rob, a second-generation American, went to Ellis Island for the first time as an adult he found it far more moving than he had expected. 'Being in the place where my grandparents first landed in America turned what had previously been a concept into an enormous sense of empathy for them. And it wasn't just the hardship or the fear that you get a sense of there, but the hope my ancestors must have had in making this journey.

'What must it have been like for them approaching American soil for the first time? It's just a generation away. I wish my grandparents had lived longer so I could ask them the questions I have now. They were part of a wave of Eastern European Jews who changed the world: they took over the movie business and brought their traditions that have become part of the American vernacular, as have the Italians and the Irish. It is extraordinary to think of the bravery and courage of these people who took this journey in order to give their descendants a better life.'

David Sye's great-grandmother is one of many who made the epic journey, although in her case the destination was England. The pogroms had become a constant threat in the region around Kiev in the Ukraine where she lived, and Cossacks were ransacking the villages at will. Her husband had already gone missing and she decided to go to her sister, who was living in Liverpool. 'She negotiated her journey by the stars, just by knowing that Liverpool was north-west of where she lived in Russia,' says David. 'She took her two young children and swam across a river that even the toughest men would not cross.' After making her way to Bremen in Germany, where she became a cook to an army commandant, she hitched a ride to Rotterdam in Holland. Here she worked as a waitress to raise enough money for a ticket on a herring boat to Hull. Finally she walked from Hull to Liverpool. Her example gave her 'number one' grandson the determination to pursue his gifts. 'One' sounded like 'vaughan' in her Russian accent, and he became the 1950s pop sensation Frankie Vaughan.

Nearly 2 million Russian Jews emigrated from Russia's murderous pogroms between 1880 and 1921. Many went to the United States, bringing new energy to the already assimilated German Jewish populations. They crowded into tenement buildings on the Lower East Side of New York with their entire families and brought a rich Yiddish culture of literature, music, poetry and theatre. Although the great majority went into the schmutter (garment) trade to survive, the trajectory of their descendants into positions of power was precipitous. And they built a vast entertainment industry in California. Even now, many Hollywood stars have Russian Jewish roots.

Immigrants formed the landscape and culture of America. It was hard-working salt-of-the-earth families from Germany, Holland and Sweden who created farming in Pennsylvania, while places like Louisiana and, especially, New Orleans find their soul – and their music and their food – from the magical symbiosis of the French, Cajun and slave families who settled there.

In the nineteenth century alone more than 50 million people left Europe for the Americas. During the Irish potato famine some 60 per cent of Ireland's population emigrated to Britain, the USA, Canada, Argentina, Australia and New Zealand. There are estimated to be nearly 100 million people of Irish descent around the world. They suffered indignities and discrimination, working hard for little money, but they overcame prejudice and became successful through sheer effort of will. And the Kennedy family became a political dynasty that achieved the highest office of all in the US.

Our ancestors emigrated for many reasons. Sometimes there was no choice: many left to escape political or religious persecution. Exile can be a painful purgatory for those unable to return to their homelands. Exiles have described it as being in a perpetual state of longing for home. My father Ferenc was one of them.

When Hungarians rose up against Soviet rule on 23 October 1956, he took part in the revolution. There was a heady moment when it looked as though the rebels might win but when the uprising was crushed a whole generation was left with a difficult choice: to stay and face Russian rule or leave their homeland and risk never returning. My uncle decided to stay but my father joined over 200,000 men, women and children who packed up what they could carry and headed for the borders.

My father managed to return on a visa in 1973 when I was nine years old. I remember the family opened their homes and their hearts for our visit: arriving late at night to find the whole village had filled the street outside my grandparents' home, we felt like royalty. My father looked shocked and grief-stricken and then warmed to the open arms of his parents and his sister Margaret. I remember a lot of tears and kissing. Decades later, as my father lay dying, he was still haunted by his decision to leave and the impact it had on his family.

Remembering the spirit of the Hungarian in exile, I recall my father saying that you had to work hard and be flexible, adapt to survive and not be afraid of picking up new knowledge and assimilating it in your workplace. 'You either integrate or you die!' said Ferenc. I have adhered to his beliefs and have always had a strong work ethic and the curiosity to keep educating myself through various mediums; these are things I am passing on to my children.

My father created his own community of Hungarians in London. They used to have boys-only weekends, playing cards and drinking until dawn, remembering their youth in the dance halls and jazz clubs of London in the early Sixties. Many of these old friends showed up for his funeral and insisted we play a classic Hungarian folk song to remember him.

The longing for home can show up in many different ways. Mbali Creazzo was born in South Africa just as apartheid was becoming entrenched. She was from a so-called 'coloured' family and they decided it was time to leave when her brother asked her father why they couldn't sit on a bench marked for white people. The entire family – including most of her mother's relatives – ended up in Crystal Palace in south London. 'We all piled into this huge broken-down old house with eleven bedrooms; there were dozens of us squashed into those rooms.'

Britain in the 1950s was a dour, depressing place still in the grip of war rationing. 'It was always grey and I remember feeling very lonely. I was literally the only person of colour in the whole playground. But mostly I remember the sense of despair in that house. Everyone was so sad. My mother held it all together. She was the strength and resilience and we were a community. But all of them died of heart attacks. I think their hearts were broken when they left South Africa.'

During the 1950s and 1960s, thousands of South Africans – white and black – went into political exile to escape the apartheid regime and avoid imprisonment or conscription. Cultural icons like Miriam Makeba, Hugh Masakela and Abdullah Ibrahim made their way to the United States where they found new audiences for the intoxicating sound of South African township jazz. Masakela tells hilarious stories of maintaining his country's cultural traditions while living in an apartment block in Manhattan, which included brewing home-made beer in the bath. 'It sounds crazy,' he said, 'and it was, but we were missing home so much and for a while it brought us back there.'

Mbali came of age during the Sixties and threw herself into the joy, colour and music of London at the time. She became a recep-tionist at the super-hot hairdressing salon, Vidal Sassoon, much to her parents' consternation – they were teachers and wanted her to become one too – and she hung out at Ronnie Scott's and other clubs in the West End, but she always felt that something was missing.

In her late forties she began training as a Dagara (West African) medicine woman, working with ancient cowrie shell divinations in which she calls on her own ancestors and the ancestors of her clients for healing. She did not go back to South Africa until she was fifty-two when, she said, her 'bones rumbled with memory'. Mbali eventually found her way home both in her spiritual practice and in her physical reconnection with her homeland.

Most of us can look back and find that we have ancestors who changed continents to fulfill their destiny. We are their progeny but how do we acknowledge them in our current lives?

Recently, I met a Persian man of about sixty who calls himself Eddie. An émigré from Iran during the era of the Shah, he moved first to London, then to Paris and finally to Los Angeles, where I would eventually meet him. A dignified character with twinkling eyes, he immediately started talking about his family. He remem-bered that, in infancy, his grandmother lulled him to sleep with stories of her mother, her grandmother and her great-grandmother and father. They seemed to him to be epic stories of love and courage. And, in turn, he did the same for his sons, as well as telling them his own dramatic story of life in exile from his homeland. 'Knowing this personal history has helped me to be secure wherever

I am,' he told me. 'I know where I come from and I know that my children know where they come from. There is value in that.'

In recent centuries, many of us have moved away from the land of our origins. This is one reason for losing the connection with our ancestors. But we have also moved away from the land itself. We have lost our sense of connection with the earth.

The connection between the land and our ancestors is familiar to all races. In rural Finland, for example, each family had a 'home tree', sacred to the family as a guardian and protector. The tree, with its branches reaching to the skies and roots going to the underworld, the world of the dead, was the means by which people could connect with their ancestors and the spirits of land. Even today farms and houses in Finland have their own special tree that is a symbol of family.

The idea of a world tree is widespread in ancient religions and mythologies, from the immense ash tree Yggdrasil in the Norse tradition to the ceiba tree of the Mesoamerican cultures, and it has been suggested that the idea of a vast tree as the world is implanted in our collective unconscious. Like the home tree in Finland, the world tree is said to connect the three worlds: the roots of the underworld, the natural world and the heavens.

Perhaps it is not surprising that we arrange our family's generational relationships on something we call a family tree. Whether or not we still live in the lands of our ancestors, our family tree can be a doorway to our reconnection with them. We can journey through our family tree into the past and travel to the land of the dead where we can meet again those who have passed on. And, as the branches of the world tree reach into heaven, so we can rediscover our spiritual and inspirational heritage. By connecting with our ancestors we can experience the light of love that flows through our family lineage, from them to us and onward to our descendants.

WHO ARE OUR ANCESTORS?

Our ancestors are all those who have died and moved on to the spirit world. They range in age from those who have just passed over to ancient elder guardian spirits. The closest are our parents or grandparents and even children who have died and to whom we are intimately connected, but we can embrace everyone in our family tree, known and unknown.

We may also call on ancestors who are connected to us through the land of our birth, our heritage or our passions. They are around us as well as within us: we see them in the topography of the land, the roads and bridges, in our great cities, in the ancient monuments that they built and the statues that honour them.

We have always honoured our ancestors. They were the invisible guides, protectors and lawmakers for the living. Modern society has lost touch with the idea that our ancestors have a place in our spiritual lives, but as we learn more about them and how to cultivate a relationship with them, we can once again call on them for support and guidance.

BIOLOGICAL ANCESTORS

Those family members connected to us by blood are our closest ancestors. Our biological family gives us our physical attributes, affects our behaviour, our health and our choices. And it is the most tangible link we have to the past. Our parents are the product of their parents and so on and, whether we like it or not, our life choices are often affected by them.

HISTORICAL ANCESTORS

Each family has ancestors who have changed the trajectory of the family tree. They may have been heroes or villains but their impact reverberates through the DNA of the family. They are ordinary people who significantly affected the world around them through their lives and work.

SPIRITUAL ANCESTORS

Our spiritual ancestors are all those who have passed on and they come in different forms. They are the spirits we honour and celebrate, and to whom we send prayers.

Firstly, there are relatives we knew in life and who wish to help us from the spirit world. They communicate with us through an intuitive voice and make us feel nurtured by being there in times of need or during family crises.

A guardian ancestor is the one who is with us when we are born and remains with us until our death and beyond. We may never have known them but they are specifically connected with us spiritually to guide and protect us and to offer their wisdom and compassion. They stand at the gateway of birth and death, watch over our children and, when we find the ways to connect with them, can lead us to our true purpose.

Finally there are those ancestors who come as a collective spiritual power to support the entire family lineage. When you connect with this power it feels like a divine force of nature making you feel supported, grounded and inspired.

SAINTS AND DEITIES

Saints are human beings remembered for their acts of goodness. Their memory reverberates through the land of their origin. Intimately connected to the land, to sacred wells, caves and rocky outcrops, saints link us with our ancient past.

A god or deity is a supreme being who guides and protects each generation. The many different religious and cultural belief systems all have their own gods and goddesses. You might feel spiritually connected to any one of the deities belonging to your family's belief system. Perhaps your family has always revered Jesus or the Virgin Mary, whereas, if you are Hindu, then you may have grown up with a connection to Lakshmi, Kali, Krishna or Ganesh. All of these deities and saints can be called upon when we need support.

TWO

Our Relationship with the Ancestors

Death does not exist
we are all immortal
and everything is immortal.
At seventeen one should not fear death,
    nor at seventy.

ARSENY TARKOVSKY

Within us all there is a silent knowledge that part of us never dies. Each and every one of us comes from lineages with customs that taught how to prepare for a good death and practised rites and rituals for honouring and remembering the ancestors. Whether we are atheists or have strong religious beliefs, we can have a spiritual relationship with our ancestors. When we are touched by death – when a loved one dies or when we have a health crisis – we are awakened to this ancestral connection.

We can communicate with our ancestors in a variety of ways:

through prayer and meditation, by celebrating them in ritual and remembering them in quiet moments. When we take time to connect with our ancestors we begin to experience the benefits of a spiritual relationship with them: we become aware of the tiny daily miracles and synchronicities, the coincidences of nature and good luck that appear as the result of their invisible guiding hands. In this chapter we explore the many ways that we can deepen our relationship with our ancestors.

The ancestor's journey: life to death to life

Our spiritual relationship with our ancestors begins with the understanding that death is a natural rite of passage. When our loved ones die, they take their place as ancestors on our family tree. From there they can communicate with us so that we feel safe and comforted in troubled times. To develop our spiritual relationship with our ancestors we first have to be willing to believe in the possibility that there is a place beyond death where our ancestors reside. It is important to have faith that they can and will communicate with us when we need their support.

Our ancestors are always with us, but it is often only when our parents or grandparents die that we open the door to our connection with them. After the death of my father I realised that from the moment of our birth we embark on a journey towards our death. I was made aware, perhaps for the first time, that life is fleeting and fragile and always at the mercy of death. This is a poignant time in our life, when the whole connection with our family tree touches us so deeply that it can lead us through the labyrinth of death into life after death. Out of darkness we are born, into darkness we shall return, dust to dust, ashes to ashes.

Our ancestors learnt to experience death as a normal part of life. In every culture there were belief systems, rituals and practices that enabled whole communities to share the experience and to assist those who were suffering from grief. These systems were ingrained in early childhood and remained powerful spiritual tools for managing their own ageing and grieving processes.

Our perception of death and its meaning begins in earliest childhood and relies on the cultural and spiritual influences that surround us as we grow up. In cultures where birth and death still take place at home there is a familiarity with death that mitigates some, if not all, of the natural fear that surrounds it. In homes throughout the developing world children will usually encounter their first death at a young age when one or both of their grandparents die. The close relationship between grandchild and grandparent means children experience a significant loss in a very real and intimate way. They learn early that death is as natural a part of life as giving birth.

In Ireland the body of the deceased is still laid out at the family home so people can come and pay their respects. Children are encouraged to view the body. With candles and rosaries, or more boisterous wakes with plenty of singing and whiskey, the dead are sent on their way and the living have an opportunity to say their final goodbyes. Laureen, an Irish friend, recalls the wake that was held for her aunt. She had died a traumatic death from alcoholism that had caused a lot of pain in the family, but during the wake her relatives were able to express themselves, to say sorry that they were unable to help her and release some of their anger and frustration.

'It was the most healing experience for all of us,' she recalls. 'As we passed round the rosary and each of us said a prayer for her soul we remembered so much about her life that we laughed and cried. We knew her spirit had left but it was the reality of seeing her body in the room that brought us all into a very real place with how we felt about her death. We really wanted her to know how much we loved her, despite everything.'

In India, the dead are burned on open funeral pyres and wrapped in simple white or golden cloth. Death is in the open and celebrated as the moment when the soul is liberated from its earthly tribulations. In the holy city of Varanasi, the banks of the Ganges are lined with corpses while near by men and women perform their personal purification rituals in the water. Life and death live side by side.

In most western culture, by contrast, death has become taboo; something hidden away in hospitals surrounded by the paraphernalia of life-saving equipment. Children are often considered too young to see a dead body or even attend a funeral.

And yet, death is one of the great truths of nature. American Indians, along with other traditional cultures who have retained a relationship to the land, have a holistic attitude towards disease and dying, so they are able to accept death as the natural outcome of being human. They view themselves and the natural world as perfect and whole, so death is simply welcomed as another transition. When we believe that the journey of the soul is eternal, that the benevolent spirit of our ancestors is real and that rebirth is a possibility, we are able to approach our death and those of our loved ones with more acceptance and less fear.

A good death makes a happy ancestor

As the physical body diminishes in every way and the essential physiological functions decline, we observe a point where the soul starts to emerge, eventually and energetically to become stronger than the physical body. This dynamic is like a quickening. A skilled soul midwife can sense this shift. A well-supported death

enables the shift from the physical to the non-physical
to become a seamless flow.

FELICITY WARNER

While birth brings us into the physical world, death takes us into the
spirit realms as we let go of life and leave our bodies. We emerge on
the other side and take our place as ancestors on the tree of life. Like
the labour of birth, the process of death can be difficult and frighten-
ing but there are ways to ease the suffering and help the dying achieve
a 'good death'. This process is so important for healing ancestral memories. The peace and compassion felt by all those who die a good death
will reverberate through the family tree just as much as the negative
experiences of those (such as my grandfather) who experienced
trauma in their death. The liberation that comes to those who feel safe
in their death changes the dynamic of grief and confusion for those left
behind, and their journey into spirit is a happy transition.

Felicity Warner is a soul midwife who has helped hundreds of
people to die and has found that a good death is a healing experience,
not just for the person in transition but also for their loved ones, their
friends and the wider community. 'I have noticed that if we can
release our ego and open our hearts it becomes easier to heal our soul
wounds, our fears of abandonment, lack of love, our anger or rage
and that helps us to have a good death. We are able to make the transition in peace.' All that is required to help someone to die, she says,
is love, comfort and compassion. And, she says, 'Having been with
hundreds of ordinary people who have crossed this sacred threshold,
I am no longer afraid of death. Death doesn't exist; it is just a shift
of energy from the physical to the non-physical.'

When my father was dying I would go to the hospital twice a day
to give him healing, allowing golden light to come through my hands
and into his body. Giving my father healing was soothing for both

of us. Words had ceased to have any meaning but we were communicating in a way that was deeper than before. The healing helped me to let him go and relieved some of my own anxiety and sadness.

Helping someone to die is one of the most valuable acts of service that we will ever perform, for the more peaceful the transition, the more quickly the spirit of those we love can find a peaceful homecoming. The last few days of life can be filled with inexpressible anxiety and pain, and it is an immeasurable gift for the dying to be with people that love them. Towards the end, people often have wonderful dreams of their childhood or deceased loved ones, or they may begin to hallucinate and have irrational thoughts. At their last breath some people call out and reach forward and then they are gone; other people simply slip away.

The difference between life and death is infinitesimal. The moment when the dying person's love and passion, the spirit that defined them, leaves and goes somewhere else is the time when the physical and the spiritual touch.

Healing for the dying

The last few days of life are the perfect time to give healing. It makes the whole painful process easier to bear and helps the dying let go. Even if someone is in a coma they can still hear and feel you, though they may not be able to respond to your presence. Many people respond to healing by sleeping peacefully for a few hours afterwards; some say it alleviates their pain and brings comfort as well as feelings of peace and tranquillity.

Massaging people's hands and feet is also a great comfort, and can be done instead of or in addition to healing. Massage gently and lightly, using a good massage oil with a few added drops of lavender or rose oil, which is luxurious and calming, or clarysage or geranium for those who are becoming agitated or anxious. You can also place a few drops of oil on their forehead or temples.

The healing practice described here is gentle and non-invasive, and when it is done with love and compassion you cannot cause harm in any way. It can be conducted at home or in hospital. Allow about twenty minutes. Always wash your hands before and after healing. Always close down after each healing. You can repeat the healing two or three times a day, if you feel it is necessary.

THE PRACTICE

- Dim the lights and light a candle; burn a little incense or heat some aromatherapy oils in a burner (if in hospital or hospice, ask permission first). Say a prayer of peace, or read the person's favourite prayer or poem out loud.

- Ask to be filled with love, peace and compassion from your God/ Spirit and the wise and unbroken ancestors. Visualise a beam of light coming down into the crown of your head and moving into your body and down into your heart. Place your hands over your heart and feel the light going into your hands. Imagine that your hands are filled with this light, then begin the healing practice.

- In most cases the dying person will be lying down. Stand behind their head or beside them and place both hands just above their head. Visualise a strong current of light coming out of the palms of your hands into their body – focus on the candle flame to help you stay focused on the light – and hold your hands there for five minutes.

- Then, still holding your hands just above the body, gently move them from the head to the throat, chest, digestive organs, pelvic area and feet. Hold your hands over each area of the body for a few minutes.

- Complete the healing by standing at the end of the bed with your hands a few inches away from the soles of their feet and visualise them connecting with the earth.

Closing down: place your hands back over your heart and imagine your heart closing from the light. Ask the wise ancestors to come and disconnect, and visualise them sending you a rainbow of light covering your aura from head to foot, to protect your energy and close down.

Death and rebirth

Death can be traumatic and sad, and grieving must be done for the physical absence of the one who has departed. Grief is a necessary cathartic process for both the living and the dead: for the living to acknowledge their loss and for the dead to experience the separation as they move into the spirit world. But death is not the end. For those who have gone it is the beginning of a new journey. And for us, the living, it is also a new journey in our relationship with them.

Humankind has created myriad ways to bury and celebrate the dead. The earliest graves have been found to contain fragments of flowers or antlers, food and tools, suggesting our ancestors have always honoured the dead and conducted burial rites, believing that the soul of the deceased lives on.

The earliest pagan cultures retained the idea of death and rebirth as a reflection of nature around them: the constant cycle of the seasons, the rhythm of the moon and stars all indicated that they too were part of the continuous spiral of life. Many early cultures buried their dead in foetal positions in the earth, as though they were, at the very moment of death, preparing for birth. This ancient and instinctive belief in life after death lies at the heart of all the rituals and ceremonies that honour our dead.

Different cultures throughout history have created their own visions of the otherworld, with their own expectations of where we go after death and what happens to the immortal soul. Hindus, Buddhists and Sikhs believe we reincarnate endlessly, living many lives until we attain enlightenment and can finally be released to heaven or nirvana. Members of the Abrahamic faiths – Jews, Christians and Muslims – believe we live only once and that when we die we are judged by God before we are allowed to enter heaven or paradise. The Pagan idea of the summerlands, a place of endless sunshine and abundance, is not so very different from the Christian idea of heaven as a place of great light where pain and suffering do not exist.

What is the otherworld, really? We can never know, unless we have been there. Near-death experiences, as well as stories told by saints, sadhus, lamas and masters who have journeyed to the otherworld, tell us that the kingdom of the dead is not a strange place, that it is actually very familiar. But to describe a non-bodily reality with words is difficult when we are used to interpreting the world through the filter of logic and science; so we have to resort to myths and maps written by others to describe the kingdoms of the dead and the idea of immortality.

It is these maps and myths – and the rituals that have been a part of world cultures for millennia – that help both the living and the dead move on. For some, like Maoris or gypsies, for example, there is a strong emphasis on rituals that ensure the dead do not come back to haunt the living; others, such as Christians or Buddhists, use prayer, meditation and praise to assist the safe passage of the soul. There is a universal belief that there is a time after death when the soul of the deceased remains unsettled as it searches for its place in the spirit realm. This period also allows time for the family and friends to gather and mourn the one who has gone. As the soul finds its place in eternity, this formal suspension of normal life enables the living to be with their loss in a sacred, loving environment and to gently grieve and let go.

In most religions there are rites and rituals that take place at this time. Jews will sit shiva for seven days after the death of a loved one, and Hindus perform kriya for eleven days: the family remains together, as they believe the spirit is still around them. An altar to the deceased, with photographs, flowers, candles and sweets, is at the centre of ritual as family members gather to sing and chant.

It is the practice of Buddhists to perform rituals for the deceased for forty-nine days after death. Tibetan Buddhists refer to this as the time of bardo – a place between life and death where the spirit seeks either a favourable rebirth or nirvana. This is thought to be a danger- ous time as the spirit reflects on the drama and disappointment of their earthly life, so mantras, prayers and offerings keep the soul on track, away from temptations that might set it off its course into the light.

Whether you uphold a religious belief or not, the actions of holding your loved one's name in your heart, lighting a candle for them on your own altar (see page 43), honouring their memory, giving money to their favourite cause or sending them love and light in your mind will all help their spirit find its way through the uncertainty after death and away from the physical plane.

We can also create our own rituals to let our loved ones go. Peggy's father died suddenly at the age of sixty. He just went to sleep one day and didn't wake up. The family was devastated and paralysed by grief. For seven months, her brother kept the ashes in his wardrobe until, finally, the family was ready to release him. 'We always knew that we wanted to put his ashes out on to the high seas because he was such a keen fisherman and my brothers had fond memories of fishing with him. So we took his boat out to the place where he would take us fishing every summer and created a sweet ritual. We lit some incense and said goodbye as we dropped his ashes into the water followed by dozens of rose petals. Then we all went swimming with him. It was a very beautiful moment of saying goodbye to him. Everything was lighter after that and there was a great sense of assistance, consolation and company. Making that spiritual connection was the most powerful consolation for me.' Now the family often dreams of him and usually he is fishing and catching lots of fish.

For members of the Spiritualist Church, psychics and mediums, as well as shamans and mystics, the spirit world is close and accessible. When Aggie Richards, an NHS nurse and natural psychic, attends funerals, she sees the living and the departed gathering together at the graveside for the burial of the newly dead. She recalls, 'I see them with no bodies, shimmering and alive with vibrant energy. Of course nobody else can see this dimension overlap, so it's as if I'm watching my own personal movie. Sometimes I see a cord connecting the dead body to the spirit and it won't leave until it is cut.'

Donna Stewart, a gifted medium at the College of Psychic Studies in London, explains death and the spirit world from the Spiritualist perspective: 'Spiritualists believe and accept that there is no death. We are all spirit, continuously evolving, mentally intelligent and sentient beings with a physical body, not the reverse. After the spirit has left the body at the point of physical death, it will be met by ancestors and spirit guides and there is an initial time of reflection on the experiences of the individual's journey so far, and readjustment of their return to this once-familiar dimension.

'The journey from here is a very personal one, in accordance with the wishes and needs of the individual spirit. Some may wish to halt their progression for a time, before continuing their journey of enlightenment and resurrection in spirit. Throughout this process ancestors may interact with family who are still living a material life. Because we are all spirit, it is possible for certain individuals with heightened mental and emotional sensitivity to have an awareness of their ancestors by way of thoughts, sensations, emotions, sounds and images that are quite separate from their own, continuing the relationship between the deceased and the living family.'

Communications from our ancestors

Life is lived with one eye to the physical world and the other to the spiritual, where your ancestors will be watching your actions and guiding your efforts.

CREDO MUTWA, SOUTH AFRICAN ZULU SHAMAN

When author Isabel Allende's beloved grandmother died, Isabel, then a young child, would imagine her spirit leaving messages for her around the big old house that her grieving grandfather had turned into a house of mourning. Her grandmother, known for her weekly séances and her ability to move the sugar bowl across the table with her mind, is still with her. 'It is her voice that I hear in meditation or when I am alone for hours writing my books. She was always about the lightness of being, about spirit and the mystery of the universe; about how we can connect spiritually without words. She really believed in telepathy, and all that comes in when I am writing.'

Isabel is constantly inspired by the voices of her ancestors. Every morning, before she begins writing, she lights a candle to her muses as she steps into the world of her imagination. 'I carry them with me as I am a traveller, an immigrant, a political exile. I have lived abroad all my life and had to start from scratch several times, so I would leave everything behind except their photographs and letters from my mother.'

It is to her daughter Paula that she turns when she needs advice about family. Paula became a mediator with the ancestors when she died at only twenty-eight. 'She is very, very present for all of us. She helps me in my relationships. I have a large tribe in California and there's always some drama going on, so before I make any decisions I talk to Paula as I know she'll always give me good advice.' For good measure, Isabel has two other adopted ancestors – Nora, her former husband's mother, whom she adored, and Hilda, the mother of her brother's first fiancée whom Isabel adopted as her *abuela* (grandmother).

Isabel doesn't consider herself particularly spiritual but, as some-one who lives from the imagination, she believes in the existence of a non-physical reality. 'There are many dimensions of reality, ener-gies, currents and influences that we can't control but which exist and can change and determine things.'

But, she says, we need to be open and present in order to be aware of these subtle realities. 'Western culture is beginning to forget our ancestors. We live always in a hurry. We live always in the noise and in the next step, so we are never present in the moment, so we never remember. But I am a writer, so my job is to remember. I spend so much time in silence and alone that I can have the luxury of having my ancestors around and remembering them.'

When we have the courage to open ourselves to the mystery that exists beyond the veil of death, we find that we all have the ability to communicate with our ancestors. As Isabel Allende says, 'It is not the spirit who decides to whom it is going to talk, it is the person who is open to the experience. You are the recipient and you create your experience.' We don't need to be a psychic, medium or clairvoyant to communicate with deceased relatives and ancestral spirits. Many people have had incredible encounters with their loved ones, seeing them both in dreams and as apparitions. Others talk of receiving symbols, such as a sweet smell of flowers, or having a 'coincidental encounter' that has the personality of the deceased written all over it.

Dawn's mother died from ovarian cancer at the age of forty-seven, when Dawn was just fourteen years old. Despite her loss, Dawn never felt that her mum was far away. 'In the first few months following her death, I would notice things about the house go missing, or be moved, or tidied away (my mum was a clean freak!) and being fourteen it was never me who tidied up. And my dad wasn't around much to tidy up either. But big heavy objects would be moved and expose dents in the carpet, or dust rings on the surfaces. It was as if she were pointing it out and telling us to pull our socks up. And there were times, when I was upset in those initial few months in the grip of teenage grief and hysteria, I would feel a cold shudder at the base of my spine and then all of a sudden a sense of calm followed, like something or somebody was giving me a big hug. I'd stop crying immediately. I'd feel instantly better.'

Even now, Dawn feels her mother's presence and talks to her all the time. And just before Dawn's best friend died of breast cancer, she said to her, 'Dawn, I've seen your mum. She's waiting for me. I'm not ready yet but she is there.'

Once we move from seeing our ancestors as people and start seeing them as spiritual beings with a broad perspective on our lives, with a vision of past and future, we can begin to trust that they know when we need their help and support.

Death and beyond: crossing the bridge

The first encounter many of us have with the presence of an ancestor is when they are dying or have just passed over. In this transitional phase – when there is a bridge between worlds as the physical self falls away and the soul prepares for the onward journey – people tell of premonitions or dreams of loved ones saying goodbye or warning them that they are about to go. They come to let us know it is okay to say goodbye.

Many of us have a strong intuition when someone we love is close to death, even when it is sudden and unexpected. Often their soul will come to us to warn us that they are dying or have just passed over into spirit. Jack remembers coming home one Friday night to find his London flat filled with an unfamiliar perfume. Suddenly his cousin popped into his mind. 'For the rest of the evening all I could think about was Michelle and the next morning I received a telephone call from my mom in South Africa to tell me that she had died in a car accident.'

Acacio remembers waking up one morning knowing that his grandmother was sick. Moments later his mother called to tell him that she had had a heart attack, so he jumped in his car and drove for six hours to be with her. 'She was always telling me fantastic stories of the "warnings" she received before anything happened to our family, so her clairaudience with me was as normal as going to school in the morning. In intensive care, I asked her who else had been to visit her (even though I knew it was just my mother). She told me that Mr Joao had been there earlier and then I knew it was time to say goodbye, as Mr Joao was one of her best friends who had died a couple of years earlier.' Acacio said goodbye and as he drove away he had a clear image of his grandmother sitting in the back, smiling at him. She died two weeks later but remains with him in spirit.

The dead go through their own separation from life and experience many different stages of coming to terms with their death, and we may find that we dream about them during this process. Just three days after my father's death, I woke up to hear him calling my name and repeating again and again, 'What happened? What happened?' For the next three nights I had identical experiences. He was so close to me that I could almost feel his breath on my face. This continued right up to the night of his funeral. It was quite disturbing because his presence was so physical and I was filled with grief.

After his funeral, my dreams changed and lightened a little. He started to show me that he could bounce up and down, that he was weightless and free. He told me that all he had to do was think of my brother and he would find himself in his house. He told me, 'There is no pain any more, no heaviness.' But it felt as though he continued to struggle. He kept asking, 'Why have I died?' He was having a traumatic post-death experience. Even though we had made him as comfortable as we could, we had not been able to help him fully process the fact that he was dying.

Then one day some months later I had another dream of him and two of his old mates from Hungary. I didn't recognise them at first because they all had a full head of hair and looked so young. My father said, 'Look, I have found my friends again.' Finally, he came to me on my birthday a year later in a dream. I was walking into my childhood church and he was sitting in one of the pews. He turned to me and smiled and said, 'I have come to wish you a happy birthday.' For the first time it felt as though he had come to terms with his death.

Raj, a computer consultant from London, vividly recalls a series of dreams he had after his mother died. For the next three months he dreamt the same dream every night, of his mother in hospital with bandages over her eyes. Each morning he woke up crying. Then one night he had another dream of his mother, this time as a young girl in a beautiful garden. 'She asked me what I said about her to the family. And I replied, "I tell them that you are dead." And she laughed. Then in the dream I saw myself looking into a pool and thinking, What are we going to do about this? And at that point I realised that I had to let her go, that she was now a fabric of God and that she wanted to move on. As Hindus we believe that death is a gift to our enlightenment. The very moment that you leave is the most important moment in the evolution of the soul.'

Just before her mother died, Lorraine's deceased father came to visit her in her dreams and described how happy he was. In the dream he showed her how they used to go dancing, her mother often wearing her favourite green high heels. Two days later her mother died. Before the funeral Lorraine got out all her old family pictures and found one of her parents: the two of them were dressed up to go out dancing and her mother was wearing a pair of green high-heeled shoes.

When we lose our loved ones, it is sometimes difficult to let go of their physical presence. We long to hear their voice or feel their touch. For some the longing is answered only by silence but for others there are real and tangible encounters that bring the comfort of knowing our relationship has not ended with death.

Briony had been very close to her grandmother, who had raised her during difficult times in her childhood, and Briony would often visit her as an adult. 'One day I asked her to come back and see me after she had died. She told me she didn't think that she could because she was agnostic and not at all spiritual and she was worried about scaring me if she succeeded. I told her to wait until she felt the moment was right.' Soon after her grandmother had passed, Briony went to take care of her house as she had promised and noticed a terrible smell. A fuse had blown and some fish in the freezer had gone off. She started to clear it out and then began to feel strange. 'I got the weirdest sensation down my spine like a tingly, fizzy feeling and a "knowing" that Grandma was there. Then I started to let silly thoughts freak me out by imagining Grandma's body floating past the door. I said out loud, "Grandma, do not show yourself as I'm not ready." I raced out of the house as fast as I could without stopping to water the plants.'

The following week Briony went to a gifted psychic who told her that her grandmother had been at the house when Briony was clearing out some 'dead food'. 'I was quite stunned she had known I'd been there and gratified my intuition had been right. Then she said to me, "There's something else Grandma wants to say . . . She said, "I don't ask much of you, Briony," (she always used this expression) "but you could at least have watered the plants!" With that I laughed and cried at the same time.'

Kerri's father Johnny Sharp died in 2000, aged eighty-six. She had grown up hearing about a documentary film he was in, made in the early 1960s, when he was a singer and entertainer in London's East End. He had never seen the film, *A Portrait of Queenie*, and Kerri's efforts to track it down had resulted in dead ends. However, in 2010, the British Film Institute released a collection of post-war documentaries on DVD and the film was on it.

'I was elated when I saw the footage for the first time,' says Kerri. 'Even though his part is a minor one, he is in the credits and shown singing and entertaining, which he loved. A few days later I dreamed I met up with Dad, typically in a pub, and told him about the film. Over a couple of pints he teased me about it, saying "What d'you want to see that old thing for?" but I could tell he was really chuffed. That was typical of him – not showing emotion but underneath you could see he was really moved. He said, "I wish I could watch it," and I told him he could watch it through my eyes. The dream was so life-like that I woke up feeling as if I'd really met him in another dimension where there was lots of love. I'm so grateful to the BFI for putting that film out, as now my mum can see it too, and of course it's very moving for her and a wonderful document to remember him by.'

There is a world of spirit that can communicate with us when we allow ourselves to cross the bridge between our day-to-day aware-ness and higher consciousness. Margaret had a pact with her mother and two sisters: whoever died first and discovered there was life after death would try to communicate it somehow to the others. 'Literally the day after my mother died, my next-door neigh-bour comes by to tell me that she dreamt of my mother and she wanted me to know that she was having a fine party with her friends. It was funny because at first I was just angry that she had gone to my next-door neighbour to relay the message and not to me. But she couldn't come to me because I was so distraught with grief. My neighbour then described three people at the party whom she would never have seen: a large man with a birthmark, who was my Uncle Warren, my mum's best friend – a woman with painted eyebrows – and another man, all of whom I had known to be part of my mother's past.'

For Margaret, sacred remembrance of her mother comes naturally. She died at home and Margaret had a powerful experience of her spirit hovering around the body. There is a strong, ongoing connection to her in the house and she is an important part of daily life. Margaret lights candles on birthdays and anniversaries and always makes sure she puts fresh flowers in the room where her mother died.

When ancestors speak to us it is often because we have a gift or leaning towards being open to spirit, so they use us as a mediator to give guidance and information to other living family members. The communication may let us know that a child is to be born into the family or that someone will soon pass into spirit: these encounters are the first step in opening up to benevolent influences. We recognise they are with us as spiritual guides and guardians.

We may dream of an ancestor for many years until, eventually, they finally disappear and move on. Emma was particularly close to her grandmother and she would often appear to Emma in her dreams after she died. Usually, the visitations were quite mundane. There might be little snippets of conversation and some questions about what was going on in the family. It got to the point where the rest of the family started asking Em, 'So, has Grandma had anything to say recently?'

One day, ten years after she had died, her grandmother came once more. This time she looked quite different: her long dark hair was tied in a braid down her back and she was dressed as though she were an American Indian, in a leather skirt that came a little above her knees. 'I remember that because in life she had this self-conscious thing about her knees and in this dream she kept pulling on her skirt to cover them.' Emma recalls that her grandmother took her hand, turned to face her, smiled and said, 'You know, darling, there is only love.' And that was the last time she saw her.

Guardian ancestors

Once the souls of our ancestors have been released they are able to take their place as an elder in the spirit world and, if they choose, they can become a guardian spirit for their descendants. These are the ones who protect and guide us, the ones to whom we can appeal for divine intervention and who protect sacred sites all over the world. They shine their light to guide us on our path towards living whole and integrated lives.

When Laura Amazzone's grandmother died it was as if the light of the world had gone out. Mammy was the only loving presence in an abusive household. Laura's only respite came when she went to stay with her grandparents. 'I would play the piano for her and she would applaud every piece. It was so sweet. And then I would spend hours reading my stories to her.' Writing and playing the piano remain two of Laura's passions; making exquisite jewellery pieces – another passion she shared with her stylish grandmother – is a third.

'But by the time I was eight I started saying, "When you die, I'm going with you. I'm not staying here without you."' And when her grandmother did eventually die, Laura was at her bedside. 'She would hold my hand and say, "You have to stay. I am going to live through you." So I promised her I would stay.'

Later, Laura went travelling in Nepal. While she was there she began experiencing overwhelming grief. 'I had promised I would stay but the truth was that I had partly gone with her and she had stayed with me. I was feeling her everywhere – seeing her, eating the foods she liked, having conversations with her. Then I started dreaming that she was having difficulty breathing. I feared I was losing my mind.'

Laura found her way to Lhamo, a female shaman certified by the

Dalai Lama, who told her that her grandmother was still in her aura because neither was willing to let the other go. 'I was not frightened but relieved. I realised that although Mammy's body was no longer there, her spirit had indeed lived on. Even though I could not see or touch her, I could feel her spiritual presence and knew we would always be connected. I had never before felt so calm and sure about any of my convictions about life after death.' After this realisation Laura did a fire puja for her grandmother and released her spirit.

Laura, now a yogini, author and teacher, continues to be inspired by her grandmother every day. She used to feel her around her cat, Gypsy, and she knows when she hears Bach's 'Prelude in C' in some random place, it is her grandmother communicating with her. Mammy is still with her – not as an attachment any more but as an elevated spirit, an elder, a guardian and inspiration. 'She was there for me when she was alive and constantly inspires me with the unconditional love of the Mother. She was the embodiment of that. She is still the light of my life and always will be.'

In helping our ancestors we are helping them in their own evolution. Once they find their rightful place in the other world then they are much more able to be present for us as our guides and guardians.

Chartwell Dutira, a musician who famously toured with Thomas Mapfumo and Blacks Unlimited and played WOMAD and Live 8, grew up playing the mbira in sacred ritual for the ancestors. He has had a close relationship with them all his life and he still keeps them close although he now lives for most of the year in the UK: 'I have my shrine at home and I ask for guidance on everything. I have been guided every step of the way.'

In Shona culture the family will visit the grave a year after someone has died, to invite the spirit to come back to the home to work for the family and their descendants. 'When they become the wise ancestors it is important for us to learn how to listen to them, to tune into that knowing and know when it's working. It's about being open-minded as to how they communicate with us. They come in many more ways than just hearing them in your mind. They can come in dreams, create coincidences, miracles. The ancestors are like the earth, wind and fire. They are everywhere and come in different forms. They are a part of what we see as our identity. It's not just my mind, it's everything; it's what makes me a human being, it makes me who I am.

Chartwell believes his life was saved by his ancestors. 'When I was a teenager Zimbabwe was still ruled by the British. I had carved a rough tattoo – the kind of thing you do at that age – of a peace symbol on my arm. White soldiers would come to the village to look for rebels, pick me up and drive me around as an example to others. My mother was scared for my safety so we had a ceremony. The shaman, possessed by the spirit of one of my ancestors, scooped up some herbal snuff, rubbed it on my arm and the tattoo completely disappeared. And so did the army. I never saw them again after that. It is something I will never forget.'

In Hawaii, the *aumakua* or family spirits are healers and sources of forgiveness; they are special ancestor spirits who do not go into the otherworld after death, but remain in the land of the living to guide and protect their families. Since the human spirit, the gods and nature are intertwined in Hawaiian cosmology, the *aumakua* may sometimes appear in the form of an animal or insect, as a rainbow, wind or mist, as well as in human form. They reside near the homes of their descendants, receiving honour and offerings while bestowing particular duties and privileges. Sometimes the family spirit has been seen sitting on the shoulder or the head of the living, appearing as an aura 'like a fire or coloured light'.

We all have ancestor guardian spirits who are always there if we need to call on them. Many cultures consider their ancestors to be like angels: messengers from God. They say that God is far too remote to be bothered by their petty everyday problems but their ancestors, still intimately involved in their family, can take messages and bring healing and abundance.

In my family, there is a clear sense of my Spanish grandfather watching over my extended family. As he has been progressively liberated from his own personal upset over his execution and the subsequent Fascist occupation, he has become more and more powerful as an ancestor.

David Sye, eldest son of 1950s crooner Frankie Vaughan and renowned British yoga teacher and musician, calls on his ancestors to help and guide him in everything he does. 'My father is with me when I need him. In a meditation once, he came to me and said, "Just because I am not in a body, it doesn't mean I am not still your father."

He continued, 'I invite my ancestors to help me before every single move I make. They cannot work without invitation but when you invite them you give them the opportunity of working through their karma and raising their spiritual energy. You allow them to evolve through you. When we worship the ancestors we are invoking a community of elders which stretches until the beginning of time. We are literally the success story that stands on the shoulders of our ancestors but we try and live in isolation and wonder why we feel so discombobulated and pissed off. It is because we are not supposed to live like that. We are not alone, we are never alone. It is so beautiful when you feel them. You just feel this great love.'

For cultures steeped in understanding of the ancestral paradigm, stories like this are as normal as communications among the living. While the dead may not be visible, they are always present watching over the family. In indigenous cultures such as the African Bushmen and Australian Aborgines, the ancestors have a human, animal and elemental aspect. Brad is half aboriginal and half English and, as a politician, he has to speak out with force and clarity about social issues. Brad told me he has a 'lightning being' as an ancestor. This ancestor is really powerful and helps him with his work, but he is not so helpful when he comes home to his wife and family. His wife Wendy confirms, 'The air positively crackles when he is around and I just can't sleep when he is near me, so we decided that we needed to ask him respectfully to recognise that I was sensitive to his power and to withdraw when Brad came home. We found out that he simply hadn't known it was a problem for me.' Wendy now honours the lightning being by placing an image of lightning on the family altar and he has now taken on the role of protector for the whole family.

Family spirits as our allies and protectors

The dead can return in various ways to reassure us, comfort us and show us their continued existence. When a relative has died I can often feel their presence around my clients. Sometimes they communicate with the living to ask forgiveness or set the record straight, but often they come simply to offer their descendant reassurance. Their presence may be marked by the smell of their perfume or aftershave, cigar smoke or another smell particularly associated with them. These trademark gestures leave the people left behind in no doubt that they are still around.

Whenever Acacio is troubled he feels his grandmother near him. 'I often feel her arms around me and, even now, her scent sometimes surrounds me when I am thinking of her. She has never come to me as some fantastic apparition but always as her very placid self with a smile, telling me that everything will be all right. She was pure love when she was alive and she is still pure love after her death.'

Evelyne only ever saw her grandfather Felix, a wealthy industrialist from Marseille, as a distant figure in a suit smoking a cigar, but she knows that she is guided and protected by him. 'When my children were young, I frequently felt his presence in our home. My son Julien said that he saw him as a cloud of smoke – so we used to call him "the smoke". I believe he is Julien's guardian angel. And I often feel him by my side, telling me things, giving me his advice. When I am having a difficult time, he comes to me in dreams. When I feel sad, lonely and puzzled I feel him as benevolent energy. It gives me a kind of feeling that I am not completely on my own.'

Ann, a lawyer on a strong spiritual path, often feels her grandmother's presence. She was an 'eccentric jazz baby' who was before her time. Self-educated and sharp as a tack, she divorced her husband in Michigan in the late 1950s and drove herself and her two daughters to California. At her funeral a relative stood up and said, 'For the longest time, grandmother was seen as just this party girl but I have realised that she was far more than that. She was this angelic force, this truly magical person who was incredibly open.'

More recently, when she moved house, Ann felt her sending her messages. 'My suitcase had been upside down in my closet for a week or more when one night I heard these weird sounds coming from the closet. Finally, I looked in and found this penny, this rock and this costume earring that my grandmother had given me. The next week, I heard some more sounds and went to the closet again and found this little needlepoint cushion with my great-grandmother's initials on it. After that I felt my whole lineage telling me they were watching out for me and that I was on the right path.'

Our ancestors will find meaningful ways to communicate with us, especially when we are going through a hard time or have serious life issues. Their communications let us know that their presence never goes away and that they do care about us and our lives. The death of my grandmother when I was nine years old was devastating for me and my cousins. She was so nurturing and gave us all so much love. Many years later, I was entering my late teens and going through a precarious time as I had just left home. I dreamt of her and could hear her calling my name 'Natalia, Natalia, *eschucame* [listen to me]'. She was trying to get my attention.

Within weeks of hearing her call my name, I attended a service at Richmond Spiritualist Church where the renowned psychic artist, Coral Polge, was demonstrating paintings of people who had passed over. The medium who was working with her called out my name and asked me if I recognised the picture that Coral had just drawn. I burst into tears as it was my grandmother. He then gave me a message from her about my future, predicting my happy marriage, my children and the many travels I would take around the world. At the time it was vital to know she was really with me. I then started to listen to my intuition and when I 'listened to her', I would get it right!

An altar is a structure on which offerings are made to a god/spirit/ancestors; also, a table used for communion to spirit and a focal point for prayer.

A shrine is a chest or cabinet for relics, sacred objects and statues; also. a place dedicated to a deity.

A nature shrine is a place in nature where we leave symbols of connection with nature, such as a candle (fire), rock or stones (earth), shells or a small fountain (water) and a windchime or feathers (air), and offerings to the earth and ancient ancestors, such as gifts of food and flowers.

A **ritual** is a celebration of life, in which we express gratitude and create a bridge between spirit and earth.

A **ceremony** gives symbolic form to our spiritual desires and helps to materialise our connection with the spirit world; generally a ritual for the entire community.

A **prayer** is a means of connecting with our personal God or deity by asking for help.

Meditation takes us into the wisdom of our unconscious mind so that we are more able to listen to our intuition and hear the voices of our ancestors.

How to communicate with the ancestors

Some people are more attuned than others to the ancestral frequency. They are the psychics, mediums and clairvoyants for whom the veil between worlds is easily lifted and who are naturally able to see or sense spirits. Usually they have been aware of their gift since they were children; sometimes it has been passed on for generations.

Lucia, from Slovakia, has always seen the dead in her dreams. 'My most vivid dream is being able to fly and finding myself on the ceiling or above the earth watching dark silhouettes walking beneath me. Often they look like soldiers and I believe I am taken to places where things have happened and spirits remain trapped on the earth.' She recognises her ability as a gift that has been in her family and continues with her nephew.

June Elleni, a psychic artist, was four years old when she first saw a ghost at a graveyard near her school. When none of the people around her validated her experience she shut down her psychic ability until she became pregnant in her thirties and it all started coming back. 'I could hear my deceased grandfather trying to talk to me and had strange experiences like falling in and out of consciousness, being unable to move, rushing through the air, hearing voices. If I'd gone to the doctor, I would have been diagnosed with some mental disorder.'

Instead she went to the Spiritualist Association of Great Britain to find ways to work with her psychic abilities. Today as an artist she uses her gift to draw ancestors and spirit guides. 'It helps people realise that consciousness survives and goes into another dimension.'

June observes that the ego of the recently deceased falls away slowly as they move towards becoming elder spirit guardians. 'I do not recommend my pictures to people who are still grieving as it can be spiritually unhealthy – they start to become co-dependent. But when the deceased are of another generation they have become spirit guides and then my pictures are useful to meditate on for inspiration and guidance. They become energy portals for people to connect with their guardian ancestors.'

Anna was extremely sensitive as a child, sensing atmospheres, knowing what people were feeling, seeing spirits. In her early twenties she found herself in a Spiritualist group where she was able to develop her psychic gifts. 'The part of the training I found the hardest was communicating with ancestors. I could see and speak to guides and communicate with Mother Earth and her kingdom but my ability to access and communicate with ancestors was blocked.

'After some time I started to connect with my client's ancestors by sensing them and I needed some way to discern who I was receiving the messages from. To my surprise my beloved deceased family stepped in to help me. During a meditation, I saw them one by one as they came forward to allow me to sense the vibration of each

ancestor. I saw my dad and my grandparents, uncles, aunts, cousins –
even the family dog and cat came in. By the time I had finished I
knew exactly who was who. I am most grateful to them for taking
the trouble to help me to understand my gifts.'

How do we sense spirits?

The cerebral cortex of the brain is divided into two hemispheres,
which are joined by a large bundle of interconnected nerve fibres
called the corpus callosum. In effect, we have two different brains
that work in harmony. The left side of the brain controls the right side
of the body and is responsible for our reason. It is predominantly
involved with logical thinking, especially in verbal and mathematical
functions. We gain our intuitive skills from the right side of the brain,
which controls the left side of the body. This hemisphere is responsible for our artistic abilities, our body image and our ability to
recognise faces. When we get a gut feeling about the answer to a
problem, it is the right side of the brain that is giving us the solution.
It is from this intuitive side of ourselves that we can see and hear the
unseen. Instead of seeing things in categories we see them as a
whole.

As small children we mostly work out of the right brain but in a
society that wants us to be logical, rational thinkers, we are often
talked out of believing that what we imagine or see is true. We start
to trust only the left brain, which really kicks in around the age of
seven when the two sides of the brain start to work together instead
of separately, as they have through the early years.

Three main psychic techniques are used to sense spirit around us and there are some simple methods for learning how to develop these techniques:

- Clairvoyance is sensing and seeing spirits, receiving information by a heightened visual awareness, seeing them as if they are physically present, sometimes seeing them as fleeting shadows out of the corner of the eyes.

- Clairsentience is the psychic sense that perceives the unseen through sensing and feeling; like a gut instinct or a physical sensation on the skin such as cold or heat, or a pressure on the back, neck or solar plexus.

- Clairaudience is sensing the deceased through hearing, as if they are talking; hearing them out loud or as telepathic thoughts.

There is a common misunderstanding that some people possess these skills and others do not. The reality is more complicated: we all have the ability to be psychic, but some people are just more awake to their powers than others. It is true that gifted psychics will have inherited their gift from an ancestor and many who are naturally gifted find that they have an ancestor of gypsy heritage, a traditional healer or psychic. It is also believed by indigenous cultures that the child who is born with the gift is there to be the communicator, a walker between the world of the ancestors and that of the living. But everyone has the potential and every individual's potential is unique.

Connecting with the ancestors

All traditional teachers say that to connect with our ancient ancestors, we must prepare ourselves by quieting the busy mind. To seek their guidance we have to raise our consciousness to a higher level. We can do this by meditation and prayer, by celebration and by remembering them and their legacy. We can all hear them if we truly listen.

The western world is largely orientated to the future and focused on achievement, so we often steer away from quiet moments of contemplation in the chaos of our daily lives, rushing from one task to the next. We have forgotten the peace that comes in being quiet and listening to a higher wisdom. To strengthen our relationships with our ancestors we have to take time out of our busy lives and create a sacred break – a peaceful moment to honour and communicate with them. Ideal times are first thing in the morning before we start our day, when dusk is falling or just before we go to bed. These quiet and sacred moments realign us with our ancestors. They help us to remember what is important and what is not.

When Mbali Creazzo wakes up every morning, the first thing she does is to tend to her ancestral shrine in a quiet corner of her bedroom. She might top up a small glass of alcohol on the shrine or bring fresh water, food or flowers. She has pictures of her parents and grandparents on her altar, and the elements are represented as well: a bowl of water and some earth from Africa, some crystals and stones to represent the mineral kingdom and a candle for the fire that represents the creative power and passion that has been passed to us from our ancestors.

'I connect with them on a daily basis,' she says. 'I am of them, and they are in me. It is because of them that I am here. I talk to them about what is happening in my life, I ask them for help and they provide me with strength, support and resilience. They are hungry for our connection and want to take care of us. Once you begin to honour them in your daily life, you enter into a mystical, magical process with them. They are clever and very cute in how they teach us to understand that there is so much more to our lives than we think. They are teaching me to trust, to let go and, as a result, I take more risks in my life because I know that I will be supported.'

In western society we very rarely look to the past and where we have come from. We see it as a long straight road leading backwards from the present. This outlook makes the past a remote and foreign place and distances people from the ancients, including our own ancestors. Traditional societies conceive the passage of time as circular rather than linear. They have an ongoing relationship with the past, and are more inclined to honour and understand the contribution that our ancestors continue to make in the world.

Constantly referring back to those who have walked before us is a way of achieving a physical and spiritual life balance. Honouring our ancestors brings a strong and confident foundation to our lives as well as a sense of continuity and protection for living family members. 'It keeps the connection alive,' says Mbali Creazzo. 'It helps us to remember that we are not alone.' We too can have an intimate relationship with our ancestors through our own personal remembrance.

When Chartwell Dutiro performs around the world he always first connects with his ancestors. 'I go to my shrine and take a little snuff and call my father's name and call on my grandfather and my great-grandfather. I ask them to open people's hearts at my performance or workshop. So then I am not alone as they come with me – I can feel them.'

His shrine is a wooden plate with some spiritual snuff from his village, where it is prepared ritualistically. He has feathers (guinea fowl and ostrich), water from a holy well and beads – white for his grandmother, red for his mother and father and black for the grand-fathers – as well as some soil from his village. 'When I know something is going to happen or my ancestors are calling me I get itchy hands and then I go to the shrine and start to clap to bring in the wisdom from the ancestors. There really is no gender. When they are coming through there is no discrimination. They will come through to communicate just as they did in life.

'When calling them I do it in a particular rota as a part of a pro-tocol. The key to summoning them and asking for what you want is to put them in order in your head. This is a bit like making a tele-phone call with the correct telephone numbers. So I call the oldest ancestor first, the oldest and the wisest, then the grandmothers, then work my way down to the most recently deceased. Then I start with my great-great-grandfather, then my grandfather down to my father.'

Chartwell confirms that the ancestors come to us in dreams and through synchronicity, but making this connection depends on us. 'This is a duty and there is no excuse that I am too busy or not feel-ing well. In this world we are always too busy and too stressed to notice these small details but it is about us anyway – it's a legacy that has been given to us from them. My advice is to pay attention in your daily life. Notice whether you receive positive calls from people, whether there is a bus strike and you felt that you should have walked to work that day, or a no-show from a new friend when you felt a whisper in your ear that they were not going to come; whether you are going to get the job or whether you are even going to win on the lottery – these are all whisperings from the ancestors.'

Ancestral altars

One of the most powerful and simple ways to continue our relationship with people who have left this world is to create an ancestral altar. This is a place of prayer, ritual and meditation, a sacred space where you can honour your ancestors and begin to feel the sacredness in your connection with them: the altar provides a powerful visible representation of the spiritual energy of the ancestors who surround and sustain us. It is the gateway between the seen and unseen realms.

CREATING AN ALTAR

Traditionally, an altar would be built in the hearth, so an old fireplace is an ideal place for an altar, if your home has one. Some people set aside a specific room for spiritual practice or make a shrine in the garden, or you can simply choose a windowsill or a side table. Choose a place where you feel comfortable and will be able to spend some quiet time alone.

Once you have chosen your space, begin to create your ancestral altar. To empower the space with ancestral memories, gather photos and mementoes of your ancestors – letters from them, heirlooms, jewellery or a book they treasured – and place them on the altar. Add things of beauty and some stones or crystals, feathers or anything else that represents the natural world and the ancestors' origins. You can include statues that represent your ancestral deities – perhaps a representation of the divine in your life: Buddha, Jesus Christ, Virgin

Mary, Krishna – or scriptures from a holy text. Use special candles and incense. The key is to create your altar with love and attention.

Purify the space and the objects. This can be done by burning incense or sage while saying prayers and gently dispersing the smoke around the altar with a feather or in a charcoal burner. Amplify the space with sound – drumming, music, chimes, bells or rattles – for a similar effect. Continue until you sense the energy in the room has lifted. Repeat the cleansing of your altar at least once a year on one of the days that we remember our dead, such as Hallowe'en, the Celtic festival of Samhain or a religious holiday that is relevant to your own beliefs.

It is important to keep your altar alive with a plant or fresh flowers and fresh water in a bowl with floating petals. Make offerings of nuts, fruit or sweets. In many countries alcohol is used, spirit to spirit, so you can add a little glass of vodka or whisky. Include anything that will help to stimulate adoration or love for your ancestors.

As well as bringing fresh flowers and water every day, make a habit of lighting a candle and talking to your ancestors: you will find that the altar begins to feel animated with powerful and loving energy from your ancestral family. This is a sacred space for you to communicate with them, a portal into the spiritual realm for sacred connection with deceased family and ancestors, a place to remember your family.

Prayer and meditation

Prayer and meditation are both means of connecting spiritually to your God or Divine Spirit and, in this case, with the ancestors. Making this connection reminds us that we are not alone: there is a greater power, a natural force that guides and protects our every move. The next step is to encourage that relationship to be a part of our daily life.

The difference between prayer and meditation is that prayer is actively participating in a sacred act of connection to the divine, in which we ask for help, protection and anything that we need, while in meditation we become still, listen and find our way to peace.

Making time to pray and meditate with our ancestors can bring remarkable grace and beauty into your life. It is an enriching process of creativity, and something that is so simple yet meaningful and reflective of the spirit within you and the spirit within your family. Ancestral spirits can help focus and bring clarity: they can see what is happening in our lives. At times of family trouble – when our children need our help or a family member is ill or there has been a death – we can call upon our ancestors to help, enlighten and support us.

Praying to the ancestors

You can pray silently or aloud in the invocation of your ancestors. 'I am seeking out the wisdom and power of the unbroken ancestors, from those known and unknown ancestors who are still close to the earth. I ask that you support and bless our family heritage.'

If you listen and watch carefully, you will witness miracles and see signs and portents to remind you that your ancestors are there to help fulfill your prayers.

THE PRACTICE

Set aside ten minutes to pray each day.

- Light a candle on your altar/shrine. Light some incense and choose a crystal to hold such as a rose quartz for love or amethyst for guidance.

- Sit, breathe and relax. Hear the hum of quiet.

- Focus on a picture of an ancestor or a deity or saint that represents the cultural and religious roots of your ancestral family. Talk to the ancestors, say a prayer, sing a sacred song, hymn, chant a mantra.

- Ask them to send healing light and love to all those within the family that need healing, protection and guidance.

- Give thanks to your ancestors, the gods and all the guiding spirits.

- Close down by saying a blessing prayer such as Amen or Blessed Be, or chant Aum. Take three deep breaths.

- Spend at least ten minutes in this quiet space, learning to be still and receive the light from your ancestral family. The main point is to connect with them and ask them for what you need at this time in your life. As your practice increases you will find it easier and easier to sit with them for longer periods of time.

Meditation to the ancestors

Meditation takes us into the wisdom of our unconscious mind, which strengthens our intuition and gives us the survival instinct to avoid putting ourselves in danger, protects us against ill health and can create luck, good fortune and opportunity.

Through meditation we can free ourselves from the constant chatter of our minds, so that we can hear the ancestors. This requires patience and perseverance and sometimes it can feel like trying to tame a wild animal, but eventually the mind will simply tire and look for a place of rest. Each time you meditate you will be able to focus for longer periods of time; even when distracted, you will return to the focal point, a place of peaceful sanctuary and personal compassion.

Meditation to connect with your ancestors

Achieving contact with your ancestors brings the physical world and the spirit world together. When you invite your spirit ancestors into your life, always ensure that you ask for the highest good, the wisest ancestors to protect and guide you.

The purpose of this meditation is to learn how to connect with your ancestors for relaxation and to learn how to receive spiritual love and support from them in your daily life.

THE PRACTICE

Meditate in the morning or evening in a quiet, warm and comfortable place. You will need ten minutes.

- Sit before your ancestral altar. Light a candle and burn some incense.

- Begin with a prayer to the ancestors, asking for guidance and help at this time. Release any fears you may have about connecting with them and remember that the spirit world is generally very wise and compassionate; it will not reveal itself to you if you are not ready for it. Learn to grow in your own time and they will treat you with care and intelligence.

- Focus on your breath and feel your breath relaxing all parts of your body. Listen to the sounds around and when thoughts arise, let them go. With each breath you take, you become more and more relaxed. Allow a short period of silence. Then call upon your ancestors with a prayer or a breath in and out and chant Aum to open your heart and mind.

- Visualise your connection with the ancestors like a beam of light coming down from them into your mind's eye and see it absorb all your thoughts – let the light open a space for you inside your mind, so you can sit and be still.

- Ask to receive the light, pure compassion and love from your ancestors. Imagine that their light and love are surrounding your body and filling your mind with peace.

- As you sit or lie still, open up your mind and body to receive this light and peace. Spend five minutes being still. Once you have gained confidence in being able to sit and receive their light, then you can begin to ask for visions, inspirations and intuitive answers to personal issues. Impressions and responses come in a number of ways, mostly sensory: you may imagine or feel their presence or compassion touch you.

- Keep an open mind and let your imagination drift, holding images that seem relevant to your questions; open a page in a spiritual or religious book that inspires you and read out the text.

- Write down any thoughts, inspiration and ideas and visions that the ancestors have given you for your day, in answer to your questions or for your family.

- Give thanks to the ancestors by saying a prayer, lighting a candle on your altar, or placing flowers or gifts on your shrine or altar.

Each time you come out of this meditation, write down what you have felt and seen. Keep a diary of this connection over a period of weeks or months. You will reflect back and see how much you have grown and been inspired by the ancestors' connection with you.

DAILY MEDITATION

Allow at least five minutes every day for daily meditation.

- Light a candle at the altar and ask the ancestors to protect and guide you and your family.

- If you have a spiritual issue or a question that is troubling you, try saying this prayer: *Ancestors of our blood, our teachers and those of our spiritual heritage, those of the land upon which we live and in which you honour your ancestors, are all honoured and called.*

- Write the question down on a piece of paper, place it on the altar and meditate on it. You may get an answer straight away, perhaps as an image or word, or you may get nothing but the answer may appear in a dream. Or you might realise you knew the answer but needed a reminder.

MEDITATION TO CONNECT WITH YOUR GUARDIAN ANCESTOR

You have begun your journey to connect with your ancestors. This simple meditation exercise will help you develop your sense of individual ancestors and how they can touch your life. It will help you to tune into an ancestor through their photograph and discover a little more about their character, the essence of who they are. Learning to connect with your guardian does take time, but conducting a meditation exercise on a daily or weekly basis keeps the doorway open between you and them.

You can pick someone you knew or use a photograph of an ancestor to whom you feel particularly drawn or who you believe to be your guardian ancestor. You may find it reassuring to begin with someone who was special to you in life. Your connection with them will be supported by the wisdom and light of a more ancient eternal guardian ancestor.

Write down their name in your notebook and, if you know any-thing about them, add a brief biography: where and when they lived, what they did for a living and their general personality characteris-tics (if you don't know anything about them it doesn't matter: you can still do this meditation exercise).

A photograph can give us a unique view into a person's life, catch-ing a moment in their history. We can draw information from a photograph about who they were, their character and what they experienced in their life.

THE PRACTICE

Give yourself twenty minutes for this exercise. Choose a moment when you know you will not be disturbed, in a room which is well lit, quiet and relaxing. Have a notebook beside you, and something to write with.

- Place the photograph on the altar and follow the instructions in the first three paragraphs of the Meditation to connect with your ancestors (above).

- Silently or aloud, ask the ancestor by name to be with you and help you gain some insight into his or her life. With your eyes closed, take two deep breaths and on the second out-breath open your eyes and look at the picture you have chosen. Spend some time staring into the photograph. Try to relax your eyes as you do this. Then allow any information to arise. You might experience a rush of information or find only silence.

- Start to ask specific questions about the person in the photograph and write down the answers you receive. Begin with your first impressions of their physical attributes: their age, their dress and the style of their hair, for example. Then, look into the emotions shown in the photograph: does the person appear to be happy or sad, angry, controlled, and so on? Try to intuit how they felt about their life and their relationship with others and ultimately, even how they died.

- Write down any thoughts and impressions about them: what they were like, when and where they were living, what they did professionally, whether they had children. Do not be afraid of making mistakes. Trust yourself: sometimes this works like a quiet whisper that needs a patient and willing ear, and the more you trust your inner voice the more you will hear it.

- When you feel that you have received as much information as you are going to get, say a prayer of thanks to your ancestor. Finish your meditation with a simple prayer: ask to be closed down, disconnect from the ancestor and ask for protection.

To further your connection with your guardian ancestor, place their photograph on your ancestral altar and light candles to them on a regular basis. Practise this meditation once a week to develop your relationship with your guardian ancestor.

THREE
Our Sacred Inheritance

'We don't know what we carry on the inside. We don't only carry genes, there are spiritual genes as well. Many people believe that we carry our ancestors' vices and their virtues. I believe that we carry their spiritual legacy as well.'

ISABEL ALLENDE

One way to continue our relationship with those who have died is to honour their legacy: who they were in life and the gifts they gave us. In telling my grandfather's story, for example, I am remembering him and continuing my relationship with his spirit.

Writer and activist Isabel Allende wrote her book *Paula* to celebrate her daughter who died, at the age of twenty-eight, due to a rare blood disorder. 'Her untimely death broke my heart. She was a graceful, spiritual young woman, the light of our family. As the first-born daughter and first-born grandchild, she held a special place in our family.'

The memoir grew out of letters she wrote to her daughter while she was in a coma and a desire to share her spirit with the world. Isabel also realised something else: 'Paula's death brought us many gifts. One was strength. I think that any mother will tell you that their worst fear is that something will happen to your child and you will not survive it. It showed me I had a strength inside that I didn't know I had. That was a great gift. Then there was the gift of love: the love of my husband, my son, my parents who were by my side no matter what. Another gift was to lose the fear of death. I am still afraid of suffering or pain which was why Paula's death was so difficult to watch, because she was suffering and we could not help her. But I am no longer afraid of death.'

During her short life Paula worked as a volunteer in poor communities in Venezuela and Spain, so Isabel set up a foundation with the mission of helping support women and girls in poor communities. 'She cared deeply for others. When in doubt, her motto was: What is the most generous thing to do? So I set up the foundation, based on her ideals of service and compassion, to continue her work.'

In this and other ways Paula has inspired her mother. 'It's a wonderful truth that things we want most in life – a sense of purpose, happiness and hope – are most easily attained by giving them to others. My most significant achievements are not my books, but the love I share with a few people, especially my family, and the ways in which I have tried to help others.

'When I was young, I often felt desperate: so much pain in the world and so little I could do to alleviate it! But now I look back at my life and feel satisfied because few days went by without at least trying. A day at a time, a person at time; in the end it adds up! In every human being there is a core of shining dignity and courage.'

And every year, on the anniversary of Paula's death, the family goes to the forest where they scattered her ashes. 'We bring a photograph of her, light candles and bring a little picnic and put flowers in the pond where we threw her ashes.' And when Isabel finishes a book she always downloads it onto a CD and buries it in the forest so Paula can take care of it.

Sometimes we are here to create our own legacy; sometimes we find ourselves becoming a legacy holder for someone who is not here to do it for themselves. When Dan Eldon died in Somalia in 1993, aged twenty-two, he had no idea what his legacy was going to be. How could he? He was too young to die. But in his short life he left behind a legacy that his mother Kathy and sister Amy recognised and – in part to deal with their grief – have turned into a global phenomenon.

Dan was born in London and moved to Africa when he was eight years old. In some ways he was an ordinary kid growing up white in Africa: privileged, curious, fun-loving. In other ways he was quite extraordinary. From a young age he was able to galvanise his imagination and his friends to make a difference in the world around him. He threw a party to raise money to get a transplant for a Maasai girl he met. He travelled to the villages outside Nairobi and brought back their jewellery to sell to his friends. He raised money to take supplies overland to a refugee camp on the border with Mozambique. It was a unique safari with a carefully chosen group of friends and colleagues, recorded by a young cameraman called Christopher Nolan. And he did all of this before he turned twenty. Then there were stints at magazines in London, New York and Los Angeles before ending up in Somalia with his camera, photographing the famine that would shock the world and the euphoria and disillusionment that was to follow US intervention.

The extraordinary thing is that he recorded all this in brilliant, soul-searching, wildly creative collages that filled seventeen A4 journals, using photos, poems, words, graphics, wax, blood, paint and cuttings. They are the witty, poignant, sexy, coming-of-age graphic storybooks of a young man with uncommon humanity, compassion and wit. In his memory Kathy and Amy have created a foundation called Creative Visions which raises money to fund creative projects that make a difference. Thanks to their efforts, the legacy of Dan Eldon's irrepressible spirit has touched millions of people around the world.

Honouring our ancestors

> Then ancestors are not dead, they are still alive. They
> are in the stones, the trees, the newborn and the rain.
>
> MALIDOMA SOMÉ

Our ancestors who were in this world on the First Day are the souls who have been and still are worshipped as divine beings by many traditional societies. They are known by many names and have been there to help us since early times. In many cultures they are held in awe and reverence: they are believed to hold the key to the wisdom of life and many of its secrets because they are closer to the origin of creation. For members of these cultures the ancestors remain a tangible, living presence, bringing guidance, protection and inspiration in everyday life. They understand themselves to be part of an eternal tapestry linking them – mind, body and soul – to their origins and to each generation that enabled them to exist. They live in the concept of the ancestral continuum and this can provide a deep sense of belonging and a powerful sense of self.

As the earth is an organic living being, it holds the memories of our ancestors within its stone, earth, air, fire, water, plants and trees. The whole of the earth recalls the births, lives and deaths of each individual that has existed and still exists on the planet. These lives are absorbed and remembered in each living breathing aspect of the earth. We find them in the land, in every stone, in running water, in the woodlands, hills and mountains. You will hear them talk to you on the wind and in the crackling of the fire, the crashing sounds of waves on pebble beaches or their gentle lapping on soft sand. So in honouring our ancestors it is important that we first honour the earth. It is no accident that we often refer to the earth as Mother Earth. After all, that is where we come from.

Inspired people all over the world treat the earth as a sacred place, bequeathed to them by their ancestors. In Peru, the native people believe that the spirits of the dead inhabit the landscape. They are known as *Tirakuna*, meaning 'the ones who watch over us'. Maori elder Pauline Tangiora says, 'The land is who we are.' She comes from the Rongomaiwhahine tribe of Aotearoa in New Zealand, close to the clan depicted in the film *Whale Rider*, and has spent decades fighting land-rights issues in New Zealand and elsewhere. 'We honour the land as sacred as it is the land of our ancestors. Women will still bury their placenta there when they have a baby and that place will always be a part of her child's existence.' She also points out that her tribe see whales as ancestors. 'They are mammals, they breed, they feed their young, just like us. So they are not a mythology, they are the descendants of the past and we see them as our protectors.'

So in honouring the ancestors, the indigenous world turns to the earth and to the elements. For them there is no separation between us and the natural world. 'Mother Earth is our original ancestor,' says Mandaza, a Zimbabwean healer who works in America. 'She is our living ancestor and we need to respect her, take excellent care of her. If you could see her as I do, you would never pollute or damage her. We need to love her – not just with our lips, but with our minds, our hearts and our hands.'

A meditation to our ancient ancestors

In their lifetimes the people revered their ancestors, went to the circles where the powerful bones lay. Now they too are dead and it is we descended from them who visit the rings, seeking our own ancestors.

AUBREY BURL

Encouraging a spiritual connection with nature is essential in order to communicate with and honour the ancestors. Whether you live in a town or in the wilderness, it is important to find a natural place that is safe and powerful at the same time: somewhere that you personally connect with, where you feel good; or, alternatively, sites where the ancient ancestors were honoured: prehistoric and Iron Age monuments, sacred wells and standing stones.

In shamanic traditions it is important to give a gift to balance your need to receive from nature and from the ancestors. Choose something to leave at your sacred spot as a gesture of respect to the earth – a plant, a crystal or flowers, perhaps. In addition, bring water and food and share it with the earth by leaving a small amount after your connection, a blanket to sit on, incense, candle, notepad and pen to write down your thoughts and experiences.

THE PRACTICE

- Spend some time walking around the site: you will be drawn to a particular spot which could be by water, high on a hill or by trees.

- Light the candle and place your gifts on the ground. Ask the protective power and wisdom within the Divine Spirit, your God to be with you. Then call upon your wise ancestors to influence your sacred ritual and give you the insights that you need.

- Lie on the ground and become aware of the points where your body touches the earth. Breathe into your whole body, consciously directing your breath into each cell and allow yourself to relax completely and surrender to the power of the earth. Then ask for guidance from your ancient ancestors and let yourself dream a little longer.

- Ask for the vision of how your ancestors lived on the earth: were they people of the sea, the mountains or deserts? As you connect, the memory of how they related to the earth will reverberate through your body, bring memories of how they related to the earth. Notice how it affects you: does it bring you peace and comfort, or strength and confidence, inspiration and ideas?

- With your eyes half closed, slowly and without hurry, begin to notice everything around you: the smells, noises, animals, objects that have fallen out of trees, plants, stones or feathers, animal bones. Find some plants, feathers, bones, stones and pieces of wood to take back with you for your shrine in dedication to your ancient ancestors.

- Say thank you to the spirits for their protection and guidance. Close down by taking several long and deep breaths and ask your ancient ancestors and guiding spirits to protect you. Complete the meditation with chiming your bells, chanting Aum or saying a prayer of thanks.

Traditional festivals

> In ancestral kinship, it is believed that the special and timeless knowledge of the old ones of the community lives on in their bones after death. The skull is thought to be the dome which houses a powerful remnant of the departed soul ... one which, if asked, can call the entire spirit of the dead person back for a time in order to be consulted. It is easy to imagine that the soul-Self lives right in the bony cathedral of the forehead, with the eyes as windows, the mouth as door and ears as the winds.
>
> CLARISSA PINKOLA ESTÉS

The ancestral elders who walked the earth thousands of years ago are evident in ancient writings and cultural stories, in myths and legends and in the traditional songs and dances of many world cultures practised today. People still flock to festivals and celebrations which are thought to bring these ancestors down to earth and where they re-enact the ancient stories in ritual, symbols, dance and drama.

Communal rituals to honour the dead remain among the most important celebrations of the year, as they reconnect the land of the living with that of the dead, appease the spirits of the ancestors who long to be remembered and so ensure the well-being of the community. From the Day of the Dead in Mexico to Dama dances in West Africa to the festival of Urabon in Japan, families and communities gather in honour of their ancestors, sometimes inviting them home with great feasts, songs and chants. These are also opportunities for the living to contemplate death: to understand through ritual that death is always present and that eternity lies beyond it.

Soul workers talk about how the souls of the dead can be drawn to ceremonies and remembrance rituals. The light of candles, the sound of familiar songs and prayers can be what they need to help them move to the light. The Vietnamese honour wandering souls in the national festival Tet Trung Nguyen, sometimes known as 'The day for the forgiveness of the lost soul'. The whole country shuts down and everyone visits graves and offers prayers and buffets of meat, rice and cakes. It is a moment when 'the living and the dead meet in thought'. There is always an opportunity for lost souls to find their way home but it is less easy if they are forgotten. These profound collective ceremonies are powerful ways to honour and celebrate the dead once more.

In the tiny village of Eyam in Derbyshire they hold celebrations every year to honour the villagers who, through their own sacrifice, held back the advance of the plague in 1665: they went into voluntary isolation after realising that the bubonic plague had struck one of their own. The vicar, William Mompesson, closed the church, people buried their dead in their own gardens and supplies were left by a well, now known as Mompesson's Well; payment was left in hollows in the stone which were filled with vinegar to purify the tainted currency. Some 260 residents, including the vicar's wife, died out of a population of 350 but they did indeed save the outlying countryside. Today, during Wakes Week, people 'dress' the well, hold an open-air service and lay a wreath on Catherine Mompesson's grave in the churchyard.

In Central and South America the Day of the Dead is perhaps the biggest celebration of the year. A synthesis of Catholic and pagan ritual, everyone congregates at graveyards and has family picnics in the company of the deceased as well as building altars in their home. No expense is spared to prepare the home altars for the arrival of the spirits of the ancestors.

Max Milligan, a photographer and author from Scotland who

lives part of the year in Peru, recalls how his experiences at the Day of the Dead changed his whole attitude to death and dying. 'They start drinking themselves into a stupor, laughing and crying as the people remember the dead. Sometimes they even open the graves and pour a drink for Mother Earth; often they exhume a skull and place it on top of the grave and pour a drink over it.'

He recalls one year when a friend of his, a local chef, was attending the celebrations with the skull of his grandmother on top of the grave. 'My friend started crying and his friends asked him why. He told them that he was missing his grandmother. So they suggested that he take her home. And he did! Now her skull is on top of his fridge with a pink ribbon around it. He is much happier now, as he can speak to her whenever he needs comfort or guidance.'

When his own father died, Max found that he was the one in the family who took responsibility for everything, as his brothers couldn't cope. 'The British attitude to death is very different from the way the South Americans deal with it. During the Day of the Dead, small children run around while the elders talked about their impending death and everyone airing their grief. It is such a healthy attitude.'

We do have our own Day of the Dead in Britain: like its South American equivalent, Hallowe'en, from All Hallows' eve, takes place on the night of 31 October. It is a Christian name for the most sacred day in the pagan Celtic calendar, Samhain (pronounced sow-hain). For Celts, this time of year used to be marked with fires and feasting to celebrate the end of the old year and the start of the new. It was a time to acknowledge the dead, as the veil between the living and the dead was said to be at its thinnest on this night. When the nights close in, we become more aware of the spirit world and more able to sense the threshold between the conscious world of the living and the twilight world of the dead.

Today, when children dress up as ghouls and ghosts, knocking on people's doors to trick or treat, they are enacting the belief that ancestral spirits would come knocking on the door of the living family and ask to be welcomed in. And the ritual of placing a candle in a pumpkin, based on the Irish folklore tale of Jack O'Lantern, is to enable those lost souls access to their family where they can make peace and restore good will.

Our ancestors would have celebrated Samhain as a sacred time of reunion between the living and the dead of the family, and it is wonderful to reclaim it for ourselves. When our children were young, our house took on a wonderful wildness at Samhain, with candles shaped like pumpkins, heaps of autumn leaves and an altar especially created for our family's ancestors. Our altar was a shrine to our beloved dead: we remembered those we knew and honoured those we had lost that particular year.

The children got involved by placing photographs and mementoes of their grandparents on the altar: their grandfather's Second World War medals, a tiny but beautifully decorated vase and antique necklace from my family in Spain, and a colourful hand-embroidered tablecloth made in Hungary, woven with multi-coloured wool like a tapestry of colour and light. We added votive candles and offerings of flowers and sweets. Our youngest son Bede adores crystals, so he placed his favourite ones around the pictures of his grandparents and great-grandparents. We took it in turns to light all the candles on the altar and then turned off the lights, allowing the flicker of the candlelight to glow against the photographs so that the faces in the pictures seemed alive. On Samhain morning, we visited the grave of my husband's father. All this was to remind our children that to honour our dead is to remember them and that makes them very happy.

When our sacred healer retreat falls on Samhain the participants join in the building of a central altar. They add their photographs, mementoes and gifts and light candles for everyone's families. The table soon fills with favourite family foods, drinks, coins, gifts, rings, objects and flower wreaths, and the atmosphere positively crackles and heaves with known and untold stories of their lives.

We teach them how to call up their ancestors, both to help heal family pain and to heal inherited sickness, either physical or spiritual, as well as introducing the knowledge that it is important to connect with a wiser, more powerful and ancient ancestor than the one who created the sickness. Finding spiritual allies is the cornerstone of wisdom in spiritual and shamanic traditions, as they bring focus, clarity and insight.

If you wish to conduct your own Samhain retreat, start on 31 October and set aside the whole day and night for honouring your ancestors. The ceremony can be conducted in your home or outside in nature. Take the opportunity to meditate about what your family needs and allow time to focus on their desires, opportunities, ambitions and talents and to reflect on the past year. Did you heal relationships that you promised to heal, and did you make the changes that you wanted to make? In any ritual or time of contemplation it is important to know where we are and where we have come from.

Julian and Carina's story

When Julian and Carina decided to marry they chose a beautiful ancient site on an island with Celtic roots and a sacred history. 'We wanted to embrace a religious ceremony as well as a sacred ritual.'

Carina's family is from Ireland and Julian's from England; they understood that this event would have a significant impact on each of their ancestral lines, both living and deceased. 'As we were bringing two lineages together, we knew how much our wedding was digging up historical memories, of our families past, as well as the emotions and expectations of our living family.'

Before the wedding ceremony they went to a stone circle near the church. 'We took gifts of local drink and food, left our offerings to honour the Celtic origins and called on our ancient ancestors to approve of our union. Inspired by the spirit of place, we approached the capstone in the centre of the circle and soaked it in Irish whiskey, as a cheer to a happy day.

'We had written two guest lists; the first approved by both families inviting all whom we loved, the second included the departed souls. As we called out their names we asked that everyone would feel loved and there would be a light over the weekend. To complete the ritual we sat at two separate standing stones. I chose the Mother Stone, which was the largest and flattest stone in the circle, and Julian took up his place opposite. With our backs against the stones we dedicated our final prayers to consecrate our marriage.'

At the wedding itself the priest agreed to honour both Catholic and Protestant religions. 'When the ceremony was complete it really felt that there was a royalty of spirits and angels and the sunlight shone through the windows as he blessed our marriage. Everything that day flowed, and we did wonder whether all the honouring and celebration of our families both living and dead contributed to the amazing big day. The weather and the mood of all family members couldn't have been any better. On our final day we went back to the stone circle to give thanks. The clouds opened to form a shape that looked like an angel. We felt truly blessed.'

Remembrance of the dead

> Remembering is an act of resurrection, each repetition
> a vital layer of mourning, in memory of those we are
> sure to meet again.
>
> NANCY COBB

The dead want to be remembered. When someone dies tragically, people often find it difficult to talk about them. The grief at their loss remains an open wound, so we sometimes stop ourselves thinking about them, leaving them feeling as though they have been forgotten.

One way that people in the modern western world connect with their ancestors is in memorials to those who have died at war. In Britain and across the Commonwealth we wear poppies on 11 November, Remembrance Day, in memory of all those who have died in and since the First World War. The poppies symbolise those brilliant red flowers that sprang up on the fields of northern France where so many shed their blood. We lay wreaths of poppies on the monuments found all over the country, where the names of the dead are listed, and we still stand in silence at the eleventh hour of the eleventh day of the eleventh month, when the war officially ended. It is a moving moment of reconnection to the past and to the sacrifice of our ancestors.

There are statues and memorials all around the world commemorating the dead, and these are powerful places. Memorials to those lost at sea or killed in natural disasters, in war, plague and famine ensure that we do not forget them.

On the tenth anniversary of 9/11, the long-awaited memorial to the dead was unveiled to the public for the first time. Two huge reflecting pools, some 30 feet deep, stand in the footprint of each of

the towers, surrounded with waterfalls cascading down glossy gran-
ite walls. The size of the pools reflects the enormity of the towers
that fell and the sound of the soothing water counteracts the noise,
fire and chaos associated with that day. Water is the curative element
in the Dagara tradition of West Africa, and reconciliation and healing ceremonies around the world use water to bring renewal. At the
first unveiling of the memorial families kissed and touched the
names of their loved ones; they laid flowers and flags and took rub-
bings of their names. Now they have a peaceful place to go where
they can be close to the ones who died.

One of the most powerful emblems of remembrance is the
Vietnam War Memorial Wall in Washington. Every year some 3 million people visit the wall, touching the names etched into the
polished black stone which was chosen because of its reflective quality: as you look at the names of the dead your image is reflected back
to you, symbolising the merging of past and present. It is so powerful because the names are listed not by rank but by their date of
death. As you walk along the wall, the ground slopes downwards and
the wall becomes higher and higher until it peaks in the middle, with
thousands of names at the time when the fighting was at its height.
There are flowers and poems here every day of the year, with people
gently sobbing at the memory of a son or a husband, brother or father
who died.

Collective memorials give us all a place to release some of our
grief and, for a moment, return the dead to us. They become sanc-
tuaries where we can feel the presence of those who have gone and
find a quiet moment to talk with them and remember them. There is
a beauty in our remembrance of the dead that alleviates at least some
of the pain of their loss and provides us with a powerful connection
to them.

At a memorial service to honour those who died during the

devastating tsunami in Asia, 273,000 petals were released from the dome of St Paul's Cathedral on to the mourners below. The petals symbolised each of the lives taken by the disaster and the flowers represented the countries affected: jasmine for Indonesia and Burma, water lilies for Sri Lanka, Thailand and Bangladesh, lotus for India, roses for the Maldives, protea for Tanzania, Kenya and Somalia, orchids for the Seychelles and hibiscus for Malaysia. For a while Buddhist ceremonies were held every year to mark the anniversary of the tragedy: on beaches throughout the countries affected, monks chanted as thousands of balloon-like lanterns were released into the night sky. The monks have now stopped the ceremony so the souls of the dead can move on and seek rebirth.

In Spain it took a generation before memorials began to be built to honour those who died in their civil war. It can take longer for countries involved in civil conflict to honour the dead as no one wants to remember what they did to each other. But it is never too late. On 5 May 2004, the Mayor, the families of the Republican victims and local people gathered in the central park in Villagarcia to erect a monument to those who died during and just after the civil war. My family was there and my cousin Maite told me how much that day helped heal their memories as they paid tribute to the fallen. The children and grandchildren of the victims told heart-breaking stories about what had happened to their loved ones and how difficult it had been to grow up without being able to speak out or honour their dead. It was the first moment they could tell their stories in public and became a cathartic experience for the entire community. Now my grandfather's memory is posthumously honoured on the monument with his full name in metallic letters alongside a long list of Republican victims who lost their lives during that tragic conflict. Flowers and gifts are still left at the foot of the monument.

On that day, my cousin read a speech that is a reflection for all those people who have died unnecessarily in civil wars and genocide, devastating the lives of their living family and descendants:

They were the forgotten ones, the defeated cast by the wayside with neither a cross nor a marker to remind them of their deaths. They were the forgotten ones, the sleepless dead; the ones mourned by their families with muffled laments and concealed grief as their teeth bit down on the cruellest dust, the ones who were exiled from their own deaths, whose death had no name. They were the dead without rest; ridiculed, stigmatised, condemned in death to a bad memory, to a memorial of nettles, to an obliteration of their traces.

They were the dead of the republic of oblivion, the forgotten of the Republic, murdered for their ideas, reviled for their ideals. They were Republicans on the platforms and in the streets. Yes, they were dead; by night they hardly shouted, but they had no wish to be shadows. I personally have read their last letters written in black ink, I have seen the paper bleed in the longest night, and I have read fond farewells, the pen strokes tinged with hope ... But their voices were faint, a tungsten filament that flickered in our dreams. Yes, I have seen that failing light and I have heard their whisper too, soft but persistent, like the call to settle an outstanding debt, like the drizzle that soaks us to the skin.

Raise us up from where we lie by walls and in ditches and bring us back.

Let us welcome this day of spring for Villagarcía because, here and now, the time to bring back those who died has begun. We are the relatives, the families and the neighbours who have inherited their memory. We are gathered together to graft new skin on to the wounds of silence. We are their replacements, the builders of new memorials and the proclaimers of a peace that does not dwell in graveyards but looks ahead, free from the shackles of oblivion. We are mindful of being the children and grandchildren of a Utopia, with the task of honouring and resurrecting their traces in order to continue in their tracks. A pledge with the past, to be carried forward to the future, has been founded here today. A shimmering mirror has been fashioned here, in which our own children, and their children, will see themselves reflected. An image has been forged here, a symbol that will revive that good memory, those good memories, in freedom and in peace.

In Santa Icía, 5 May 2004

Gustavo Pernas Cora (translated by Manda Denton)

How to remember and honour the dead

It is important to pray for our ancestors and loved ones, to meditate about them and offer remembrance to them on a regular basis. Remembering and honouring those who have died before us keeps the communication between the physical and the spiritual open.

Honouring your ancestors when you are alone is a ritual; when you share the honouring with other people it becomes a ceremony. Traditional ceremonies have been handed down over generations and are conducted in an exact way, but you can follow your own rituals and ceremonies. Making a memorial for someone at their favourite place or taking flowers to family graves are both simple honouring rituals.

THE PRACTICE

Choose a special place to honour them: the place where they are buried or their ashes were scattered, the place where they died, a communal memorial site or somewhere they particularly liked. There is a sense of celebration and ceremony when others are involved, and you may wish to share this time with other family members and friends. Good times to do this are at anniversaries, birthdays or cultural holidays.

- Always begin by lighting a candle on your ancestral altar, next to the photograph of the person you are remembering, and honour the person by bringing flowers or plants to offer them light and love.

- Choose a theme that represents their passion in life and find ways of expressing it. You can pray, read out a poem, play a special song or piece of music, and leave a gift or something that has been made specially for them, such as their favourite foods.

When you pray for someone who was close to you, you can, if you wish, extend the embrace of your compassion to include other dead people in your prayers: the victims of atrocities, wars, disasters and famines, or those who have recently died. You can even pray for people who died years ago, such as your grandparents and long-dead members of your family.

Freeing our ancestors

> As the deceased takes on spirit essence, he or she may
> get snagged into thinking of himself or herself still as
> a person. Thus, the deceased may begin to intrude into
> the business of the living in a way that can constitute
> a serious nuisance.
>
> MALIDOMA SOMÉ

Death is a rite of passage for both the living and the dead. Funeral
rites and rituals help us let go of our loved ones and help their souls
to ascend into their own spiritual evolution. But sometimes their
spirits remain stuck, or attached, to the material world. They are then
known as earthbound spirits.

Often the bond of love is so strong that they cannot let go of their
family. Maybe they have unfinished business or guilt about what
they did in their lives. Or they have lost their way and stay in the
material comfort of the family home. Spirits are especially prone to
becoming earthbound when their death has been violent, tragic or
sudden. The soul, usually unprepared for its death, needs to reconcile itself to the life it has lost before it is able to move on. When
spirits are earthbound, they are trapped by their life and death, endlessly reliving their pain.

The dead have no physical voice so they find other ways to communicate, by projecting their thoughts and emotions on to family
members until they are heard and their story can be told. I have
found that many of my clients' physical, emotional and psychological problems can be linked to an ancestor who has become attached
to them. If you are suffering from nightmares or have inexplicable
phobias or anxieties, you could be reliving the trauma or sudden

death of a particular ancestor. Children are often the first to react to the presence of an ancestor who is not at peace, and in most cases the spirit will attach themselves to those in the family who are the most sensitive, who carry a certain light that they cling to.

There are many ways to help these lost souls. The first is by recognising that they are trapped and finding ways to release them: it might take a simple prayer or ritual, or it may require a professional shaman or soul rescuer. My husband Terry O'Sullivan has worked as a soul rescuer for some thirty years, helping to rescue spirits from buildings – stately homes, flats, offices, health centres, even churches – or land sites like graveyards and battlegrounds. He is able to walk between the world of the physical and the dimensions of spirit. As a 'walker between worlds' he connects with lost souls and helps them find their way beyond the veil of death and into their own personal resurrection. Meanwhile, as a mediator between the living and the dead, my gift is in being able to communicate with those in the otherworld, giving them a voice to express unresolved issues with their living family. This helps both the living and the dead to let go and forgive the past and reassures the dead that they are remembered. Usually, there is a process of letting go, forgiveness, reconciliation and release.

When Tania sought my guidance, she had been having dreams of being held down or paralysed at night and it frightened her. 'I'd wake to see a dark shadow sitting beside me, I could never go back to sleep, I thought my mind was playing tricks on me, but I started recognising it was actually a "presence".'

We explored her family tree to try and find the source of her trouble. I sensed that there was someone in her family who had died in a fire. Tania realised this to be the cause of the haunting. 'When I lived in Africa my grandfather was killed instantly in an accidental gas explosion in his kitchen. I was five years old but I only found out about it when I was fifteen.' Now she had identified him as the one who had been haunting her she was able to help him move on. 'We were able to release him into God's loving light and sever our bond. I felt that he needed to be noticed so that he could let go and move on.'

The effects of an ancestral attachment come and go. With some personal awareness, you may be able to discern when your emotions are being affected by an ancestor rather than by your own personal mood swings. There is a sense of 'not quite feeling like yourself', feeling sluggish and unfocused, a heavy feeling in the head or heart. The triggers could be returning to the family home, a death or birth of a family member or the anniversary of a death.

Anna sought my advice as she was suffering from anxiety and had problems making long-term decisions. Anna was afraid to stay in London, afraid to move, afraid to get pregnant, afraid to miss the chance of having children. She was paralysed by fear and indecision.

As our consultation progressed, I saw an ancestor: a man in his fifties, quite short with dark hair and wearing a darkish suit with silver buttons. She identified him as her grandfather and explained that he had been shot dead by her grandmother as he was having an affair. Her grandfather explained that he was attached to her because he needed forgiveness. Together we prayed that he would find his way. Then I saw her grandmother appear. She exuded an aura of love and compassion and forgiveness. They embraced and as they did so the room seemed to lighten up. Anna's mood completely shifted. She was elated for weeks afterwards and has come to see her life in a completely different way.

Sometimes an earthbound spirit's influence can be quite malev-olent, affecting more than one generation and causing depression, distress and other mental health issues. Margaret brought her daugh-ter Annabel to see me as she was suffering from acute depression and anxiety, had started to self-harm and often talked about suicide. A previously easy-going character, Annabel had suddenly started manifesting these symptoms at the age of twelve. Margaret knew they had a troubled family history, as her father had killed her mother in a fit of rage and then killed himself in prison. In her childhood she had suffered nightmares of being chased by a presence she couldn't see. Now Annabel was having the same nightmares.

We began working on releasing Annabel's ties to her grandfather, as it was clear he was projecting his guilt and trauma onto his grand-daughter. We placed his photograph on an altar and said a prayer to the ancestors to come and help us release him and create a space for forgiveness and love. Then Margaret and Annabel visited his grave and left flowers and a letter for him and said a prayer of forgiveness. Annabel has been free of her nightmares ever since and has now enrolled in college to complete her studies. Giving a voice to ances-tors who have died in tragic circumstances can liberate their spirit as well as heal descendants who may be carrying the 'sins of the fathers'. And the power of forgiveness is amazing: so much can be repaired by a simple act.

In her divinations, Mbali Creazzo also finds that ancestors attach themselves to her clients following an unexpected or tragic death. 'The soul has been suddenly taken away and there is this notion that it continues to look for a way home and in the process they tend to want to "stick" (literally) to people who are still living here. For the living it has the effect of feeling being pulled back, of not wanting to be here, of feeling half here and half in the other world, so it can stop them moving forward, it can keep them stuck and cause this sense of feeling lost or confused.'

Often the family ceases talking about the one who died. Everyone quietly buries their grief and, as a result, the spirit feels excluded. 'By doing a simple ritual you can release them. I usually tell people to go to the ocean, use their own words, make their own ritual to cut those threads to the person who is clinging to them. In this way, you are able release the person and heal your grief but you are also bringing them back by honouring and remembering them.'

So it is important to honour and remember them but also important to let go. In some families grief keeps the spirit of their loved ones in an earthbound state for years. Gary died in a motorbike accident when he was twenty. His girlfriend Sally came to see me because even ten years after Gary's death neither she nor his family could let him go. Now she was about to get married and needed to release her heart connection to him.

During her healing session Gary told me about his fears of leaving his family and in particular his mother, who talked to him every day as if he was still alive. He blamed himself for his accident and believed his mother had never recovered from it. He was willing to let go but just didn't know how, so Sally and I prayed for his ancestors to come and shine a light into his heart. We asked them to show him that there was so much more for him in the afterlife, so that when he next revisited his family he would have experienced the joy of the release from his old life and be able to return with a new perspective to help his family. He did leave and went with his maternal grandfather.

The following day Sally called me to say that she felt so much lighter and that she had had a dream about him, laughing and smiling, and showing off his shiny new motorbike. His mother had the same dream and for the first time she felt at peace and knew he was now happy.

Anadru, a Maori healer, says that it is important to bury or release the ashes of your deceased loved one after they have died, as part of the protocol of rituals after death to help souls move on. He told me, 'A woman came to see me because she was experiencing psychic phenomena in her home and she could hear, see and smell her late husband. I went to visit her home and immediately noticed a pot on the mantelpiece and asked what it contained. "It's my husband's ashes," she said. I explained that the ashes would be the cause of the husband's manifestation, as he had not been buried in consecrated ground. When she later buried the ashes, the phenomena stopped and she felt that her husband was now at peace.'

When someone takes their own life it can leave their family grieving and in despair at the loss. The taboos around suicide and the unresolved questions and guilt can leave both the living and dead in a place of limbo.

Angela Watkins is a medium known in the Spiritualist Church for her gift of clairvoyance. As a Spiritualist she has a unique perspective on life after death, but even so it was a great shock for her when her daughter Amanda took her own life in 2005. She was only thirty-six years old but she had developed ME (myalgic encephalomyelitis) and felt unable to cope with the physical and emotional demands of her illness.

Some months later, when Angela was in the middle of a church meeting, she saw Amanda sitting on a piano stool. Amanda told her that she had come to help her work with families who were suffering from the death of loved ones in tragic ways, particularly suicide. For Angela this was a godsend, as it is these families who find it hardest to recover. 'Amanda now guides me towards those who've lost a child to suicide or tragic death, so I can give them an explanation and understanding to help them heal their loss. It brings them so much peace. As the spirits communicate with them they stop searching for the reasons why their loved one took their own life and begin to accept it was their choice. Amanda has her feet firmly established in both worlds and I am so happy that I can still communicate with her. I feel that Amanda has found her peace by helping others like herself to connect with their families.'

When someone we love dies suddenly our grief is traumatic. The dead, too, experience anxiety and grief. They want to prevent the tears and sadness of their loved ones and can remain in limbo until they realise that they stop the progress of their own journey into spirit by remaining too close. We can heal their trauma and our own by sending prayers and honouring them, by remembering who they were and reminding them by these actions how much we still love them and, finally, by letting them go.

Healing the family home

Some earthbound spirits remain attached to the place they believe is still their home, their refuge. When I was just nineteen, I spent the summer in Patak, Hungary in the ancestral home where my family has lived for at least six generations. Every night I slept badly, waking up with a heavy feeling in my chest as if someone was trying to suffocate me. One night I awoke and saw the face of a woman in her fifties with fair hair and blue eyes. In the morning my aunt told me about a great-aunt who had been unhappy and died in her fifties. She had married a gypsy and was forced out when the family refused to accept their relationship. She returned when her husband died but never felt that she was accepted. She died alone just after the outbreak of war.

The next night I decided to call on my ancient Hungarian ancestors to help her. I felt her enter the bedroom and this time she stared intently at me and then floated away. A peace came upon the room and she never returned. I placed some flowers on her grave in a gesture of peace and forgiveness.

Unsettled ancestral spirits attached to the family home can pervade the atmosphere with their own personality and unhappiness and slowly their issues can affect the family or anyone else in the house, including later owners. After Nicola's grandfather died her family wanted to rent out his beautiful converted farmhouse. There was no obvious reason why it shouldn't have rented out immediately but the house lay empty for months. Nicola convinced her family to call on a soul rescuer. Terry, my husband, found her grandfather's spirit still in the house, which had been left exactly as it was when he was alive, with all his pictures on the walls, antiques and Victorian furniture. He made it obvious that he was unwilling to let his house be rented to strangers. Nicola and Terry conducted a ritual to heal the house, clearing the atmosphere so it could be more welcoming to another family. During the ritual Nicola's grandmother appeared and said that she had come to call him away. Three weeks later the house was rented to a couple with young children.

When you rent or buy a property, you will be inheriting the atmosphere of the people who lived there before you. As you decorate and make the place your own, the atmosphere dissipates and your personality takes over. But what if there has been a murder, traumatic death or a suicide?

A friend had moved into an apartment in an old building in Glasgow. She loved it but had a problem sleeping. Every night, she would wake up with a start and feel a dark presence close to her side of the bed. She kept telling her boyfriend that something was going on but he dismissed her concern. One night it was particularly bad and she decided she had to do something about it. Coincidentally, when she went for her morning coffee her waiter said, 'You live in that apartment block on the corner, don't you?' She nodded and he said, 'You know there was a murder there?'

She called her landlord and he confirmed that an old lady had been killed when she surprised a burglar. 'I knew immediately that it was her who was waking me up every night.' When she went back to the apartment, she lit a candle and talked to her ghost. 'I told her I had found out what had happened to her and how sorry I was. But I asked her to please find peace and move on. And, to my amazement, she did. I was never woken up in the middle of the night again.'

Find out the history of a place before you buy it. There are so many old buildings that have been turned into hotels or conference centres and housing developments that were previously used as a hospital or reformatory or may have other past issues associated with them.

A spa hotel in Taunton called us to help them as they were experiencing all manner of disturbances. They had had several complaints from guests about random psychic phenomena: doors would suddenly open, there were sounds of running, beds would shake. There had been occasions when breakfast staff would turn up and all the knives and forks had been moved around the tables.

The manager told us that it been a manor house owned by the same family for several generations. He went to find a photograph taken during the Edwardian period, with the family and estate workers outside the front of the house. He pointed out that a number of the estate workers' children had died of tuberculosis and been buried on the property in unmarked graves. At exactly midnight that night we heard a rattling of the door as it appeared to open, followed by the noise of several children running down the corridor. Then a teacup I had left on the floor was sent flying.

When Terry called out to them, a young boy aged around thirteen presented himself as the eldest, and leader, of a group of eight children. Terry found out they were very upset as their parents had all disappeared, leaving them alone with strangers. Terry explained that their parents were not there because they had died and moved into the light of the spirit world. He suggested that they could follow them there. The phenomena at the hotel completely disappeared and never returned.

Rescuing the ancestors from places of mass suffering

> In every culture we have ancestors who were violent or died violently. We have the power to heal the yester-year wounds of our ancestors for if we continue to allow them to operate among us, there will be no peace. We are cleansing the old history. The old ways. The old wounds. So we can bring in a better history and create a world of peace.
>
> MANDAZA KANDEMWA,
> A TRADITIONAL HEALER FROM ZIMBABWE

When we moved to Somerset we became very aware that it is a part of the country steeped in historical memories and the souls of the dead. In 1685 the Battle of Sedgemoor between the rebel army of the Duke of Monmouth and the royal army led by Lord Faversham ended in the ruthless slaughter of wounded rebels: more than a thousand were killed. Monmouth was captured and executed in London. Ghosts of horsemen have been seen galloping over the battleground. Disembodied voices are heard and the ghostly figure of Monmouth is said to re-enact his attempted escape every year. This atmosphere has not yet left the landscape.

In any landscape where there has been mass slaughter or suffering there is a shadow imprinted on the land and the presence of the earthbound souls is still palpable. The violence and suddenness of death makes it very difficult for any soul to make a successful spiritual transition. In Gettysburg, Pennsylvania, where 50,000 soldiers were killed in the bloodiest battle of the American Civil War, the memory of death still lies over the town. It is one of the most haunted towns in America, with daily sightings of ghosts and other paranormal activity. The battlefields of the Somme brood with the memories of some 300,000 Allied and German troops who were killed there in 1916, and there is something dark about Glencoe in Scotland, the site of the infamous massacre of the MacDonald clan by their hosts, the Campbells, in violation of strict clan protocols. It is as if the land absorbs the energy of such tragedies as it absorbs the blood of the dead.

The death camps of Auschwitz, Belsen and Treblinka, and notorious prisons such as Alcatraz in San Francisco and Tuol Slong in Phnom Penh in Cambodia, still hold an eerie atmosphere of death, pain and suffering. Countless souls died here, and they are places that need healing and honouring just as much as the deceased. Priests, shamans and healers spend their lives visiting these sites and helping these souls to move on.

When we were invited to attend a wedding in Ireland, I began having dreams of crying babies and infants, mothers calling out names of children with such distress that it filled me with a deep sense of loss. When we arrived at the hotel I found an old cemetery alongside the hotel grounds. There was a plaque by the site of a mass grave for women and children who had died during in the 1845 potato famine. This was the source of my dreams. Some spirits were still trapped and the atmosphere felt like a grey cloud, a heavy grieving shadow encircling the graveyard.

Further along the ancient wall we found a Holy Well, where people had tied ribbons, left trinkets and gifts for a statue of the Virgin Mary. I prayed to the Virgin for help and guidance to deliver those children. I threaded my way around the cemetery and lit a candle to call all souls to the light. It was such a beautiful afternoon and, as I finished my prayers, light broke through the clouds and I was bathed in warm sunshine. I was joined by my husband Terry, who called on Archangel Michael to deliver these souls out of their suffering to be led away into the light. We completed the rescue by giving thanks at the Holy Well.

That night, on the eve of the wedding, I had an amazing dream: the hotel had been refurbished in bright colours of light blue and gold and the landscape had also changed, with a rainbow spreading itself as a bridge across all the fields and hills. I woke up elated, knowing that the souls had truly been rescued.

Releasing the ancestors

Three years had passed since my father died but I still noticed him around and I kept dreaming about him. Often they were happy dreams but I knew he was unsettled. I felt the restlessness of his spirit was connected to the way he had left his family and never truly returned.

That summer we decided to take some of his ashes back to his Hungarian homeland, back to the family home in the northern region close to the Slovakian border. I wanted my father to find peace and I also believed that his family needed closure. I always thought that his sister Margaret had never got over the fact that my father had left during the uprising and rarely returned home.

When I was a child, my grandfather came to see us in London. With his warm face, bright blue eyes and grey hair, he had gentleness about him but also great sadness. He had survived two wars, been imprisoned as a prisoner of war, lost two brothers in the Second World War, and survived an invasion and then an occupation. The

Soviets had confiscated his farm and left the family with very little. And then he had lost his only son.

His family was devastated when my father left. They didn't even know whether he was alive or dead until, ten years later, he called to tell them that he had married and had two children. Just before he died he called his sister to say that he was coming home in a final gesture of making peace with his family. But he died before he was able to make the journey.

That summer we flew to Hungary with my father's ashes tucked in my handbag. It was twenty years since I had seen his family but my early memories of Hungary – generosity, affection, love, food and the most beautiful countryside – all flooded back.

The family met us in Budapest. My cousin Tibi had hardly changed, my aunt Margaret just looked a little older and smaller, and my uncle still had his ruddy farmer's complexion and affectionate manner. As I was embraced by them, I felt my grief for my father overwhelming me and I had to hide my tears because I knew that if I cried, I couldn't stop. During that short stay the dreams of my father got stronger and more physical, as if I could reach out and touch him.

On the day we chose to bury his ashes the whole family came: my aunt and uncle, all our cousins, their children and my family of five. We made quite a procession of united Kovacs heading uphill to the village graveyard where our family had gravestones reaching back at least five generations. Margaret placed my father's ashes inside her parents' grave, we said a prayer of remembrance for my father and then placed flowers and gifts.

I so wanted my aunt to forgive my father for what had happened and at that moment I felt that she had. The hearts of my family had begun to heal. We lit candles, drank the local liquor Palinka and told stories. As the sky lit up with a thousand stars we wished for his liberation and for his whole family to be at peace, so the damage caused by the Russian occupation would be healed. I made peace for us as a family and for my father.

The next morning I was bedridden. I rested, I slept, I found peace. A few days later we returned to England with a lighter heart and an empty box. My connection with his family since coming home is close: we are regularly in touch and our children communicate with each other on social networks and delight at the prospect of another visit. It has been five years since his death and I know in my heart he is at peace and so are we.

Stairway to heaven

This is an exercise that I use when there is an unquiet soul in the family. Begin with a willing and open mind to open a gateway between the seen and unseen realms. The ancestors will know that they can come and assist you in this powerful representation of spiritual energy.

Many earthbound spirits find it difficult to move, and we can be a mediator between them and the wise guardian ancestors who come to rescue them. With the use of prayer we can be the conduit between the physical and spiritual worlds.

THE PRACTICE

• Begin by lighting a candle on your ancestral altar. Place a religious or spiritual deity on your altar to represent your family's religious or spiritual beliefs, and place family photographs on the altar, with the photograph or name of the unquiet soul/s.

- Say a prayer to your God/Divine Spirit and ask that the ancient ancestors come and assist in the rescuing of unquiet soul/s. The prayer should include a statement: 'We realise that the dead are not dead and for the light and peace of spirit to come and rescue them.'

- Call out the names of the unquiet soul/s and ask them to come now. Ask them to embrace their past and forgive. Tell them that you have forgiven them and that they can let go – they can go into spirit to be with their loved ones.

- Visualise a stairway to heaven – at the top of the staircase are the many faces of their loved ones calling them into the light. Feel the light surround them and send them up the staircase. When they have all gone you will sense a quiet peace and with that the heavens will close and the staircase will vanish.

- Pray to your God/Divine Spirit and the ancient ancestors, give thanks and ask for protection and release.

Healing the Family Tree

We continue the chain of generations and knowingly or not, willingly or unwillingly, we pay the debts of the past. As long as we have not cleared the slate, an 'invisible loyalty' impels us to repeat a moment of incredible joy or unbearable sorrow, an injustice or a tragic death. Or its echo.

ANNE ANCELIN SCHUTZENBERGER

Our ancestors both enhance and obstruct our personal destiny. By reaching into our inheritance we can study generational patterns and make a break with the past. We can heal our family tree.

Ancestral influences affect us in many different ways; ancient memories, transmitted into our subconscious, can affect us deeply, preventing us from becoming who we truly are. Ancestors who have died suddenly or traumatically send reverberations through the family tree, so each generation suffers the pain of their loss. And our collective generational legacy, the history of wars and civil wars, holocaust, genocide and famine, lies in our bones, haunting our memory and affecting our life force.

Whether you are shadowed by ancestral memories or affected by collective family trauma, there are various ways to help heal the past. In this chapter we will explore ways to heal our family tree so we are able to liberate ourselves, reach for our highest potential and bequeath a better legacy to our descendants.

The power of forgiveness

> Our freedom comes when we realise that in order to thrive and survive and to break out of our prisons of resentments and hate, we do not have another choice but to forgive and live our true purposes – fully alive. It is in this way that we become good ancestors.
>
> ALEXANDRA ASSEILY

When people come to see me for a consultation they often meet the one ancestor they least expect – the family member who caused the most pain or dysfunction, who is usually the one who holds the key to healing the past. Aware they have caused suffering, they come to seek forgiveness and offer advice and support. When we are able to forgive them from our hearts, we begin a new cycle in our families. Without forgiveness we are destined to stay stuck in the old stories, and the suffering continues.

Sonia's family had a long line of dysfunctional, self-hating women and she herself was physically and emotionally abused. After years of therapy, she came to a point of forgiveness. 'I came to see them as women with their own trauma and limited understanding. I cannot judge or disown them because of how they behaved. I am their DNA, so to completely love myself I had to love them unconditionally too. So I prayed for all of those women and I forgave them. Now I feel it has been transcended and healed for me and my children.'

Janine, from a similarly abusive background, turned to her religion for ways to forgive her grandmother and her mother. 'The female line of my family is very damaged and it has taken me a while to understand what we carry from our family and what we have to let go and stop carrying for them. Although I was the dutiful daughter, and still am, I have learnt that it is okay not to like your parents. However, as a Reform Jew, forgiveness is very much part of my daily practice. And during Rosh Hashanah, our New Year, we spend ten days of repentance culminating in Yom Kippur, the Day of Atonement, when we name things in our hearts that we could have done better and are encouraged to seek forgiveness from anyone who we have upset or from God – ideally both. I know now that I have forgiven my mother and my grandmother. I found a way through my family's pain by giving my love and compassion to others, in my work as a charity director and by nurturing my relationship with my sister and her family as well as my friends.'

Forgiveness doesn't mean that we forget the abuse or the transgression; it doesn't mean we give up our right to justice, but it does mean that we can let it go. You forgive, the saying goes, and you find out that the person in prison was you. Bishop Desmond Tutu, chairman of the Truth and Reconciliation Commission in South Africa, says, 'To forgive is not just to be altruistic. It is the best form of self-interest. When I talk of forgiveness I mean the belief that you can come out the other side a better person, a better person than the one being consumed by anger and hatred. Remaining in that state locks you in a state of victimhood, making you almost dependent on the perpetrator. If you can find it in yourself to forgive then you are no longer chained to the perpetrator. You can move on, and you can even help the perpetrator to become a better person too.'

When our lives have been thwarted by abuse or alcoholism, violence or crushing limitations or lack of self-belief, it is difficult to forgive those in our family who have been the cause. But our relationship with them does not end when they die; the situation can continue to haunt us or we can learn to forgive. This is a long process. 'It took years of personal healing work,' says Mona. 'Then one day I had enough of feeling wounded and agitated around my family so I said to myself, I'm just going to have to forgive myself and forgive them. And that was it. Now there isn't even scar tissue; it's as if it didn't even happen.' Sometimes it is just in the letting go, by being the one that understands the family legacy, that we can stop the cycle of trauma, abuse or hatred within the family tree. There are always those in the family who understand the responsibility of the family legacy and, with the help of the ancestors, they can stop the cycle of trauma, abuse or hatred within the family tree.

Julia began her healing journey by discovering what had happened to her great-grandmother. She had spent ten years systematically healing her family line, using a mix of meditation, prayer and invocation, but was unsure whether it was working. Then one night she was awakened by her great-grandmother, who told her that she was furious with her great-grandfather. He had been abusive to her in their marriage and had then gambled away his inheritance and run off with the maid, leaving her penniless with seven children to feed.

'After I found out about her story I understood that this is what I had been carrying all my life: a deep unconscious hatred of men and a real fear of poverty. This transformed my relationship with my husband and also with money. As I healed this memory I was able to love him so much more. And I stopped being afraid of losing everything. But I was still finding it very difficult to forgive my great-grandfather for what he did to her.

'Then I went to see a medium and she channelled my great-grandmother, who told me, "When I became destitute I would pray that my grandchildren and great-grandchildren would be liberated and free, and that is why you've always had this search for the truth and love within you." This was her inheritance to me. If she hadn't lost everything our generation would have had money but no soul. I was actually able to thank my grandfather for that and was able to completely forgive him. That seemed to be the conclusion of healing the male aspects of my family.'

'The ancestors are always looking for peace and reconciliation,' says Mbali Creazzo, 'and our connection to them gives us the opportunity for complete redemption: to redeem the broken connections and heal the wounding. It is easier for us to not forgive, as this helps us avoid change and transformation. But if we stay stuck in the story "he did this" or "she did that" we continue the wounding of our legacy. The ancestors give us a huge opportunity to heal our wounds through forgiveness, peace and reconciliation.'

Alexandra Asseily was inspired to create a Garden of Forgiveness in Beirut as one way to stop the inter-generational cycles of pain and violence that had plunged Lebanon into a sudden and grievous civil war when she was living there.

'It began with a question: is there a contract between the living and the dead? Do the living hold the dead and the dead hold the living? I felt with every fibre in my being that the answer to my question was an unqualified yes. We carry an unconscious contract with the dead that makes us continue their grievances and carry on their wars. Because if we don't then we feel that we are betraying them. The only thing that I know that releases that contract is forgiveness. The only thing that will bring us closer to lasting peace is a different way of facing our memories than we have used before. This requires leveraging the formidable power of the human heart in order to forgive.'

This, she says, happens on a personal and global level. When we forgive the tyrant – whether that is our father, the Catholic Church, the dictator or the family, country or tribe who suppressed us – we release ourselves and them from the cycle of hatred that otherwise continues into perpetuity. 'Peace starts within us but we cannot get to peace without forgiveness – even when we think we have done nothing wrong we all need to ask ourselves, What is my responsibility for peace in the world? And that starts with forgiving.'

Currently, a simplified version of an ancient Hawaiian system of reconciliation and forgiveness known as Ho'oponopono – literally meaning 'to make things right' – is becoming very popular in the USA and elsewhere. The simple prayer 'I am sorry, please forgive me, thank you, I love you' is used as a cleansing mantra, whereby we accept full responsibility for whatever it is in us that has caused conflict in or around us, past or present; this in turn liberates others.

Lei'ohu, a Hawaiian kahuna, agrees the tradition is much more complex, involving ritual, invocation and symbolism, but she believes it has reached the outer world for a reason: in its essence it restores harmony in our relationships with family and with our ancestors; and that is needed in the world right now.

Healing the family

Because we are not encumbered by spending most of our time just trying to survive, we are at a point in our evolution where we finally have the time to devote to breaking free from these negative patterns so they are not passed down to those who follow after us. The negative legacy from our predecessors can have a staggering effect on our emotions especially if they are not addressed.

DENISE LINN

When we are born into a family we enter a complex matrix of relationships. We are part of an extended family that includes the living, the dead and the not yet born. Yet in the West the concept of the nuclear family, the dispersion of the family unit and a disconnection in our relationship with our elders, as well as with those who have died, have limited our understanding of the influences affecting us. We enter the world trying to make it on our own while carrying the hidden unconscious burdens of our ancestors that we ignore until they begin to affect our life, our relationships, our work and our health.

Ancestral history can cloud our personal awareness of our true self and shadows from childhood can dictate our feelings and the way we think, just as our parents were clouded by their parents and so on. Descendants of victims of war and persecution talk about the depression and anxiety that undermine their lives. People discover that their own mental problems can be linked to a great-grandmother who was incarcerated in an asylum. Or we may find that the alcoholism that seems to be plaguing the family began with a grandfather who returned from war. Healing needs to reach beyond the veil of death to where our ancestors may still be restricted by their own emotional trauma. Previous generations of our family had no time to soul search as we do, so by helping to heal their pain in us we are helping to heal them too.

The beginning of healing is recognising that you are not the one causing the pain, but that there is very likely a link to disturbances in your family tree.

A woman suffering from sexual problems recalled stories of a maiden aunt who died bitter and alone because she was prevented by her father from marrying the man she loved. A man finding it impossible to earn and save money discovered his grandfather was an inveterate gambler who pitched the family into poverty. Both were blocks they had inherited from the past and needed to heal if they were to go forward in their lives. When there has been a chaotic or unhappy ancestor, or a generation in trauma, it will affect following generations. Trauma, sudden death, secrets and shame create blocks in the natural flow of love from our ancestors. Ancestral healing removes those blocks, so balance and equilibrium can return to the whole family lineage.

Sometimes when we absorb an ancestor's life story into our own it is difficult to know where their story ends and ours begins because it has been with us for so long. However, we are of a generation that has the ability, the time and the resources to transmute these experiences and heal the past. When we choose to do that, we are in effect healing generations past and future. Once we see it we can deal with it.

During our Sacred Healer weekends many clients have vivid dreams that bring plenty of tears and laughter. Sometimes they are about the ancestor who needs healing; there may be a traumatic memory that lies embedded in the nervous system of the descendants. Or ancestors themselves come to bring healing, messages and comfort. Death is not the end we imagine it to be; it is simply a portal through which all of us travel and through which we can return if we wish, to complete unfinished family business. The important thing to realise is that our relationship with our ancestors continues to change and evolve, encouraging forgiveness and new understanding of who we are and our role in the world. By the end of the retreat, there is a palpable sense of connection, a profound sharing of spirit and information.

This is a first step on a journey that might take a lifetime to complete. It is a journey of rediscovering through a gentle process of separating ourselves from the issues that plague our family. The journey is one of evolution and each step, each ritual, each healing, each prayer, takes us closer and closer to the lives our ancestors are dreaming for us.

Healing addiction

> Fidelity to ancestors which has become unconscious and invisible governs us. It is important to make it visible, to become aware of it, to understand what impels us and possibly see if we may not have to reframe this loyalty in order to become free again to live our own lives.
>
> **ANNE ANCELIN SCHUTZENBERGER**

Addiction is often triggered by a loss or trauma that creates an open emotional or psychological wound. Coming to terms with the loss can heal the memory and lead to recovery. By looking within our family and asking where the wound began, we can begin to heal it.

Annie comes from a family of addicts and she herself was an alcoholic for many years. While she used AA to control her compulsion she wanted to find the source of the sickness in her family. 'I had a sense our addictive behaviour was more emotional than just genetic and I wanted to find out what that was about.'

She discovered that she came from a long line of disconnection from parents: absent parents, emotionally absent fathers and grandfathers and great-grandfathers who were shell-shocked from both the First and Second World Wars. 'I realised we were all responding in our own different ways to abandonment. I had been emulating my parents as a way of connecting. My core symptoms and addictions were due to an emotional entanglement with my ancestors, and I had a choice whether or not to continue carrying this for the family. From that moment my life changed.'

Pia Mellody, founder of Mellody House in Arizona, considers addiction to be the result of deep-seated emotional trauma. The treatment programme works with addicts to release unresolved trauma – or frozen memory – lodged in the limbic system, making it almost impossible for them to deal with the normal ups and downs of life without self-medicating. By unravelling emotions from childhood and exploring learnt behaviour and belief systems such as guilt, fear, anger and sadness, and then releasing these emotions from the body, the addicts are able to recalibrate themselves to a new way of being.

Dr Carl Jung described the addict's craving as a 'thirst for oneness' and suggests that at its root is the very human longing for completion and wholeness. This gives us the opportunity to observe our addictions without judgement. There is no shame or blame involved. If we can witness our struggle to overcome our addictions with a compassionate knowledge of where those addictions began, we can find a way to heal them. And when we understand that our compulsions are a human inheritance that causes us to seek 'oneness' or wholeness, we can begin to find other avenues that will bring us to our goal without compromising our health and sanity.

Malidoma Somé, a West African shaman and writer renowned for bringing the wisdom of ancestral teachings to the modern world, believes the loss of ritual in western culture is one cause of soaring addiction levels. In performing rituals people are able to connect to spirit and the ancestors, dissolving the sense of separation and loss. As he points out, when western colonisers around the world removed indigenous rights to practise ritual, alcoholism became endemic. And it is the return of ritual that is healing these communities once more.

Addiction begins with a routine or habit that works to anaes-thetise or sedate uncomfortable emotions; when we learn to sit with those emotions we begin to replace bad habits and compulsions with good ones. Most addicts are strong but wounded characters who, with the right support, can change the triggers for their addictions and develop the self-confidence to replace addiction with more pos-itive personal rituals. As we answer our inner call to become more of who we are, heal our wounds and feed our inner light, we can eventually chase away the shadows. Yoga, meditation and walks in nature all soothe the craving that otherwise pushes us towards unhealthy compulsions.

Agatha Rodgers, a therapist in Devon, helps addicts identify, release, heal and transform wounds using ritual and ceremony. 'One of the rituals starts with a "no send letter" directed to their pain, their late parent or loved one, to their ancestor, their abuser, to themselves. Then they read the letter out loud to me or to a group. Once they have finished I leave a time of silence to allow their energy to settle and then ask them how they feel. They often talk about the process, images, thoughts and feelings. We then proceed to the "release" stage in which they can burn or bury the letter.

'They are asked to speak their prayer for release and healing and to give an offering to their ancestral guardians: this can be flowers, candles, plants, water from a sacred place, crystal or stone to express their gratitude. Before they go they are asked to share with the group the way forward for them.

'These rituals can be very simple or very ceremonially charged and they help unlock the energy and transform it into something creative, positive, respectful and healing. With ritual and talking therapy we are giving back to the wounded the power to heal themselves.'

Healing infertility

> Women are almost never told how their family history,
> beliefs and emotions affect their fertility. Knowing this
> information can be very empowering.
>
> **DR CHRISTIANE NORTHRUP**

One of the most obvious ways in which generational family patterning is passed down is through the conception and bearing of children, but we are experiencing an epidemic of infertility in the West. In America in 2005 the number of women aged 15–44 with an impaired ability to have children reached 6.1 million and the number of couples who were infertile reached over 2 million. As much as 20 per cent of this infertility is 'unexplained', meaning there are no medical reasons why people cannot conceive.

There is increasing evidence that our fertility is impeded as much by our minds as by our bodies. And, while science is coming up with myriad ways to overcome fertility issues with IVF, surrogacy and hormone treatment, sometimes the healing of our ideas and fears around childbirth is the most effective approach.

In healing memories of guilt, shame and fear around illegitimacy, abortion, miscarriages and adoptions in previous generations, we find the keys to unlock our fertility. Our mothers, grandmothers and great-grandmothers may have had inner conflicts about having children that are now being reflected in our own body's reluctance to conceive. What happened to them? What were they like as mothers? Was their birthing experience traumatic? Did they even want children? When we look into the circumstances of our mothers and their mothers we may find real psychological conflicts regarding parenthood.

Sjanie Hugo Wurlitzer, a clinical hypnotherapist specialising in fertility, recalls a woman who was struggling to conceive. In hypnosis she had a strong recall of a young woman falling pregnant out of wedlock and leaving her religious family to have the baby. She was alone with this young child, really poor and deeply ashamed. 'The story made no sense to my client but she did some research and found out that her grandmother, who had grown up in rural Ireland, had indeed fallen pregnant out of wedlock and had the baby in very shameful conditions. My client was carrying around the subconscious belief that it was not okay to conceive without being married. So she got married to her partner and she fell pregnant some time after that.'

Fertility issues can bring us to explore deeply where we come from – and that is a good thing. Often the women who come to see me have unresolved issues with their mothers who have been absent, critical, competitive or undemonstrative. They come from a long line of unmothered mothers and consequently are unsure that they themselves are capable of being loving mothers. There is a deep distrust of the feminine and a sense of feeling unsafe in the world that counteracts their conscious desire to conceive. Making the connection between your emotions, your family and your ancestors can open the way to conception.

I had a client who had miscarried many times but was sure she was supposed to have children. We talked about her parents and she confessed that her mother was very unmotherly while her father was an alcoholic. We worked on cutting the ties with her parents and even burnt a picture of her with her parents, in which she was looking a little forlorn and very unhappy. That night she bled with an early period and the following month she fell pregnant with her daughter.

Men, too, are affected by familial relationships that block fertility. They lack confidence in their sexuality and their masculinity when they have distant or judgemental fathers. Jeffrey and his wife, Jenny, came to do a number of ancestral workshops with us to heal his relationship with his family. They were trying to have a baby and were having difficulties. His grandfather was a military man with no intimacy or affection for his family, so his father learnt to be cold and unaffectionate with his own children. Jeffrey did some personal releasing of family ties, embracing the warmth and affection from his mother's family to balance the coldness and absentness from his father's side. Jenny was also able to release her family ties and feels grounded in her body for the first time. They are now trying for a baby allopathically as they feel they have released the toxic emotions from their families.

Conceiving is a receptive act: by removing blocks and past emotional shadows we send the message to our bodies that we are ready to receive. Sometimes it helps to call on the ancient ones for help. Gill knew she was going to have fertility problems. Her grandmother had her mother at forty-one after a great struggle and her mother had problems getting pregnant with her and her brother. They have all inherited polycystic ovary syndrome and, when Gill started to try to get pregnant, she found it difficult.

She looked back to her great-great-grandmother who had had eight grandchildren. She did a great deal of meditation work with a healer as she sought to break what she believed to be an inherited issue with infertility. After many months and some sessions with an acupuncturist, she fell pregnant. 'I called on my great-great-grandmother to help us have children and I do believe that she heard our prayers and now we are having a child. We opened some champagne and poured her a glass as well.'

 In the ancient and traditional worlds, fertility was one of the prime gifts that we hoped to receive from our ancestors. They were the ones who would gift us a child from the spirit realms.

It was for the fertility of our fields and our bodies that we brought offerings and prayers to ancestral shrines and did rituals on special days in the hope of pregnancy. Infertility was sometimes seen as a curse or a sign of displeasure in the ancestral realms. However, the extended family did much to mitigate the personal loss. Children belong to the community when aunts, uncles, mothers and fathers are interchangeable and responsible for rearing and loving children.

In Hawaii the sense of family or *ohana* does not just rely on blood ties. Children are loved and esteemed whatever the circumstances of their birth. When one woman cannot conceive, another will gift her a child. It is a time-honoured tradition known as *hanai*: the child knows its biological genealogy but is connected with an extended family relationship. Today, western rules around adoption are messing with the traditional system with paperwork and legal issues but *hanai* still exists and is a way to keep the community intact and spread *aloha* or love throughout all family lineages. Something similar to *hanai* is happening in the West with surrogacy and adoption and gifting of children through family systems.

Tai and Vivien had almost given up hope of having a child after ten years of trying and several cycles of IVF. At forty-two, Viv was considering adoption but her heart was breaking at the bureaucracy of it all. On Tai's birthday, a friend treated them to a full-blown Druid ritual at Stonehenge. The Druids considered the transit of comet Hale-Bopp as an opportunity to manifest their 'seed dreams'. 'We were a little intimidated as Druids weren't really our thing,' says Viv, 'but at one point the Chief Druid took his staff and pointed it into the sky and said, "I open the dragon's path to the comet," and, like a Disney film, the clouds parted and we could see the comet with its trail blazing. All the Druids started laughing and clapping. They were as surprised as we were and that helped us feel a little more comfortable.'

Tai and Viv approached the altar at the centre and quietly asked for a child and picked up one of the crystals on the altar. 'And that was it. We didn't think anything more about it although we did plant the crystal in a part of our garden that a Feng Shui book had told us was the baby corner.'

Three weeks later they received a late-night phone call from Viv's sister in America to say she was pregnant and wanted to gift her child to Tai and Viv. Viv went to bed for a week in shock. 'I was worried that my sister didn't really understand the implications and I had made a mistake when I had cried for a baby from the universe and forgotten to specify that the baby come from my own loins.' But she went out to America and realised that her sister – with four children and a farm with dozens of horses – did understand what she was doing.

When Kim John Sequoia was born at home in Kentucky, Viv caught him and Tai cut the umbilical cord. For three weeks her sister pumped her own milk so Viv could feed him. When they returned to England with their new baby they were a legal precedent for 'international kinship adoption' but some kind of magic cut through the red tape and Kim has basked in the love of two families ever since, bringing special love to his grandmother in England.

When infertility is not resolved by this kind of miracle, it brings on a crisis when we discover we may not have children and continue our lineage. Then we are called to heal the trauma of infertility. Even though the stigma has lessened for infertile women in the West, there is still a great deal of pressure from parents who want grand-children, and society in general is geared more naturally to families with children. In any case, it's the end of the line and brings with it a daunting sense of finality, knowing that we are leaving no biological legacy.

But making sense of our heritage and learning from it can bring a sense of empowerment and wholeness. When we understand why we might have made the unconscious choice to be child-free, we can get on with creating a legacy that is more than just biological. We can provide an inspiration and legacy to countless nephews and nieces, godchildren and the children of friends just by following our calling and showing them what is possible. When we think in terms of an extended family or a community, we can see how child-free adults can teach things that parents, burdened by responsibility, cannot. Their legacy can be an extension of familial love, beyond blood and duty, that offers inspiration and mentorship to the next generation.

Healing sudden or tragic death

> When the dead are honoured, respected and remembered, and their fate is acknowledged, they are a source of healing and benevolence. When judged, rejected or forgotten, they can cast a shadow on the living.
>
> **ROSIE ANSON, FAMILY THERAPIST**

Sudden and tragic death naturally shakes a family tree to its core. When family members are killed or die too young, when there is violence involved or some shame or blame attached, a shadow is cast over the living descendants and may still affect them four or five generations further down the family tree.

The problem occurs when the person is not honoured, respected or grieved. Sometimes, the pain is so great that the family buries its grief and stops talking about the one who has died or else there is some secret shame which means that person 'disappears' from the family tree: their name is never mentioned and the circumstances of their death are quietly forgotten.

Helen's story illustrates the multi-generational effects of traumatic and sudden death on both sides of her family. Helen was four years old when her younger sister Julia died from meningitis. The trauma rocked her family. 'My mother was so heartbroken that she couldn't speak, nor could my father. They got the local vicar to tell my older brother, myself and my twin brother and we all knew we shouldn't mention her name.'

Her father, a doctor, blamed himself for her death and began to drink to dull the pain. Helen's parents divorced when she was ten years old, but she always managed to maintain a relationship with her father even when her other siblings could not. She discovered the reasons behind her father's inability to grieve and how the patterns had replicated over the generations.

When her father was ten years old, his beloved elder brother had been killed in the Battle of Britain. 'Uncle Roy had already lasted nine months in the air force and, on his last leave, he came to my father's room and told him that he didn't think he would be coming home. "My number is up," he said. And it was.'

Helen's grandmother disappeared into her grief. 'I think my father completely closed down after that. My elder brother was exactly the same age when Julia died and my parents did the same thing.'

Helen was also curious about her mother's side of the family, as she had no heirlooms and there were no stories. She just knew there was something unspoken that had happened. Eventually, she pestered her father so much that he told her, at sixteen, the story buried in the family's past. Joy, her maternal grandmother, had accidentally killed her grandfather when she discovered he was having an affair. She had pushed him and he had fallen heavily and knocked his head. She was devastated. The local GP sympathetically recorded a natural death but she was never able to forgive herself and committed suicide three years later.

Helen's three uncles were all traumatically marked by the family history. The eldest married and moved to a remote island off the coast of Scotland where he farmed his land with his wife. Then his wife caught him having an affair and shot him. (Helen was told that he died in a sailing accident and only found out the truth later.) Another uncle became an alcoholic and died young. The third travelled the world but there are rumours he had a disabled child that he left behind. 'I realised that our family had just built trauma on more trauma on both sides.'

The healing began in Helen's generation. Her elder brother overcame alcohol addiction and, after her father died, all her siblings were able to heal their relationship. 'At my father's wake, which I held at my house, there was a moment when my three brothers just held each other in the middle of the room. It was a very healing thing for our family.'

And the healing continues into the next generation. Their younger brother was born after Julia's death and so escaped much of the trauma his elder siblings suffered. He is the only one in the family to have had children. He has two daughters. The first is called Katherine Julia. 'My mother always calls her by both names. It is the first time she has spoken my sister's name since she died.' And the second daughter is called Maria Joy. The name fell from Helen's lips one evening in a quite unconscious way and then she realised that this was the name of her maternal grandmother. In this way both Julia and Joy have been restored to the family. 'The circle is complete and our family feels, if not exactly whole, then certainly more complete and healed.'

According to Malidoma Somé, sudden death can itself be the result of our disassociation from our ancestors. 'A culture in which people die accidentally is said to have a dysfunctional relationship with its ancestors. It's the ancestors' way of attracting attention to something crucial to everyone's well-being. You know that the ancestors are healed when things begin to change dramatically for the better around you. The healed ancestors will bring health, prosperity and a sense of intimate connection that is unparalleled.'

All our ancestors need to be honoured, especially those who died tragically, so the trauma of their death does not continue to disturb the family. When we honour those who have died, we restore them to the family tree, thereby rendering whole that which was broken and bringing many unseen blessings for the family.

During a guided meditation a young girl introduced herself to Nicola as her grandmother's sister and said she was there to help Nicola from the spirit world. Nicola immediately felt her sense of deep grief. 'I've been here all this time,' the girl said, 'but no one can see me.' Her name was Hilda and she had died in 1916 when she was just fourteen from a blood disorder, probably leukaemia. Nicola's grandmother was left an only child and Hilda was barely mentioned again. When Nicola mentioned Hilda to her mother, she gave her an old locket with a tiny picture of a young girl in a school uniform and a lock of hair. While her great-grandmother had never talked about Hilda, she obviously held her close to her heart all her life.

Now the locket is on Nicola's altar and she regularly lights a candle to Hilda, thanking her for her presence and honouring her memory. 'The reconnection is indescribable,' she says. 'It was as if some light returned to my life and the family in general.'

The missing children

The invisible presence of children who have died young, been miscarried, aborted or stillborn can affect other members of the family tree in various different ways. Stories tell of unborn children and miscarried siblings becoming guardian angels to watch over their living family, but sometimes the pain and grief are buried and they are forgotten.

It wasn't until I went to Spain to do research on our family history that I found out my grandmother had given birth to a son called Adolphino who died in infancy. My cousin recalls that when she was young my grandmother would ask her to get her a bottle of wine on the anniversary of her son's death. She would drink it all and then cry and call out his name as she sobbed. She never mentioned him till the following anniversary. She lived at a time when it was essential to keep all your emotions close to your heart and she carried these sorrows to her grave, hopefully to be reconciled with both her beloved husband and child.

Hidden grief coils around the family tree, touching us in unknown ways until it is addressed and healed. Sjanie, a hypnotherapist specialising in fertility, recalls an Indian client who came to her because she was finding it difficult to fall pregnant. When Sjanie asked what needed to be healed in order for her client to conceive, her client felt this surge of inexplicable grief. Sjanie discovered that she had been the second-born in her family and that the first-born had died. Her mother was still grieving when she fell pregnant and even gave her newborn the same name as the one who had died. 'My client realised she was carrying her mother's grief and that she needed to heal this grief before she could have her own child.'

Miscarriages and stillborn babies are quickly forgotten when another child arrives and yet the child that follows feels both the unspoken grief of their parents as well as the unaccountable loss of their sibling. There is often an unconscious pull from the otherworld or a survivor's guilt that leaves an incoherent sense of incompleteness. It also changes the dynamic in the family as birth order is capsized.

Angel Gutierrez, a curandero (healer) from Spain and herself a child who followed an in-utero death, explains: 'The soul who came before you clears the way for you and you remain soul-connected with them. My brother died because my mother was taking something that she shouldn't have been, which meant that my life was saved. Sometimes our soul makes an agreement with the soul of the one that died to live for them but this doesn't work well if we are to live our own lives.' She recommends that the child soul be acknowledged and honoured; that we give thanks but also ask them to please let us go and live our own lives. If children who have died are remembered and honoured, their invisible presence can become a source of love and inspiration instead of grief and pain.

When Sara Connell, author, speaker and coach, suffered the stillborn death of her twin baby boys she could not at first find a way to 'make it meaningful' as the counsellor had told her and her husband to do. It wasn't until they travelled to Mexico and came across the concept of 'babies that pass through', in clay statues of mothers with tiny babies on their shoulders and a full-sized baby in their arms, that they began to find a way to honour their loss. 'Upon returning home, I found similar honourings of "babies that pass through" in other cultures; including a sect of Judaism in Israel who believed the children that depart through miscarriage or stillbirth became spiritual intercessors for the family.'

A month later they were ready to bury their twins. 'Bill and I had the babies' bodies cremated, and placed their ashes on an antique table in our garden facing west, the direction of the ancestors, in tiny twin urns, one with seagulls and the other with grey baby dolphins, painted on to shiny blue bases of sturdy steel. I saw the urns often as I passed from the garden to the kitchen, from the bedroom to the landing, my arms full of laundry. They were not emblems of sadness to me, but touchstones of hope, calling me to keep faith-filled and committed during the three long-feeling years ahead when I would have my son. My dialogues with them were not appeasements for the children we still longed for but words of thanks for the gifts they had been to us; the experience of being pregnant, the deep love we already shared.'

Some time later she was reminded of the idea of departed babies as intercessors when she heard another mother's story. Her first child, a boy, had died late in pregnancy and she had gone on to have a little girl. When her daughter was three, the mother heard her babbling to someone in the nursery. When she asked who she was speaking to, she said, 'My brother. He always comes to talk to me at naptime.' The mother had never mentioned that she'd had a baby boy who died and couldn't fathom how her daughter knew.

Honouring the passing of such a short life brings some sweetness to the bitterness of loss. In giving them a name we give them an earthly presence and, in all of this, we make life and death sacred. The soul who touched our lives even briefly brought with it many gifts and lessons, and performing a ritual or ceremony makes it easier to recover from the trauma of the loss, to grieve and let go.

Jamie's story

Jamie is American and living in Spain with her partner. She was brought up as a Catholic but has an open-minded view of religion and all things spiritual. The couple were delighted when she fell pregnant. 'I already had a daughter by my first husband, but my partner had never had children so we were over the moon. But I asked Rafa not to tell anyone, which is very unlike me, so maybe I already knew something might be wrong.'

All her regular tests kept showing small inconsistencies and she began to find this worrying. Then her worst fears were confirmed when doctors told her that her baby boy had mosaic turner syndrome, a chromosomal defect that is extremely rare in boys – one in a million, she was told. She was given a choice: to keep the baby or to end the pregnancy, which by now would entail giving birth. She chose the latter but before she went through with the procedure we did some healing work together.

'I had no doubts that my son was not ready to come to earth. He needed to be somewhere else. We began working on releasing him and I felt him leave my body. I saw my deceased grandmother come with two angels and lift him up in her arms. Natalia told me she saw him throwing hundreds of rose petals as he went, and as he moved out of me to the light I felt soft pink petals drop down on my body as my tears started to flow. I felt empty but happier and softer about this whole painful process. It was so peaceful and calm.'

The next week she went to have the procedure. 'It was difficult but healing had made all the difference. And I was able to hold and touch him.' Jamie and her sister organised a cremation and blessing for her child, and in her heart she named him Mateo.

'It has taken a year to heal and still now I have my moments. A friend gave me a special Indian bracelet, blessed by her guru, to protect myself and Mateo as we were healing. When it broke, she told me that I had to put it in the sea. Six months later it fell off and one day driving home, I decided to go to this beach that is very special for us. I threw the bracelet into the sea and said a little prayer. When I got home, Rafa said to me, "Do you know what day it is today? It is Saint Mateo Day!"'

'I will always remember Mateo. Even today when I think of him I see rose petals and rainbows. He is one in a million.'

Increasingly, doctors and hospitals are becoming more sensitive to parents who are called to bury their unborn child and will release the remains to them for that purpose. In cemeteries there are now designated areas filled with tiny graves, surrounded by wind chimes, little teddy bears and toys. The Catholic Church offers blessings for parents after a miscarriage and will conduct a mass in memory of your child at your request.

In Japan a modern Buddhist ritual for miscarried, aborted or stillborn children has become popular. The child is known as *mizuko* ('water child' or 'unseeing child') and in the ritual known as *mizuki kuyo*, an offering is made to Jizo, a bodhisattva who protects children. In the ritual the *mizuko* is encouraged to move on so that it may find another way to be born.

The Vietnamese honour all lost souls and teach that children lost due to miscarriage, abortion or illness are also our children. Thus parents who have two living children and who have also had an abortion and a miscarriage will say that they have four children.

When their second son was seven months old, Song's wife Lan had a miscarriage. For a long time afterwards, Lan had a disturbing dream that a baby appeared and pushed their infant son from her breast to get milk for himself. She finally consulted a family elder. He had not known about the miscarriage but asked if she had had another baby. The elder instructed Song and Lan to build a tomb and an altar for their baby. They constructed the altar underneath their window and the dreams immediately ceased. Now they put sweets and toys on the altar, as with other family tombs, and they always say they have three sons, not just the two who are living.

Giving back what is not ours to carry

> The traumas that befell your grandparents weigh on your heart like lead. Transformation comes when you hear the ancestors. They mourn when you carry their burdens and rejoice when you receive their gifts.
>
> **DR DAN BOOTH COHEN**

Angel had always been angry. She was angry at everything, railing at traffic, rude people and family members who she felt had slighted her. 'And once I got mad,' she laughs, 'I got really mad and I wouldn't let go of it. It had become a bit of an issue.'

She realised she had taken on her father's anger and that it wasn't really part of her own spirit. Brought up in Spain, his family had suffered during the civil war. He was angry with the world and often took it out on her. Eventually he died of a heart attack. 'Anger is one way to get connected through love but as a child I just absorbed it and pretty soon it became a part of me.'

She calls the way children take on the issues of their parents, both consciously and unconsciously, 'blind love'. 'We want to belong, we want to be loved and so we will take it on. Either we become like them or we try to alleviate whatever it is that is making them suffer: we absorb it into ourselves and then when we become adults we lay claim to it as if it were our own. But that is not the case.'

A curandero (healer) and therapist, Angel performed rituals for her father. 'Basically I gave it back. I said to him, "This is yours, not mine, and I don't want to carry it any more."' Her anger calmed down and her weight issues also faded. 'I had thought that my weight issues were related to my mother but, once I healed the anger issue with my father, my relationship to food completely changed.'

'We cannot change what happened but we can give it back,' says Angel. 'We can see that nothing has been done on purpose but the pain is passed on to us none the less. By seeing it clearly and without judgement we are able to give it back and then move on with our lives.'

Cami Walker, a *New York Times* bestselling author, began suffering from mental illness when she was twelve years old. She has been diagnosed with a whole gamut of different variations: bipolar disorder, clinical depression, anxiety. When she was twenty-three she began to research her family history. 'I realised that I came from this long line of mad women. It totally made sense to me as I had always felt that my mental health issues were not mine – and now I had found out where they came from.'

When Cami was trying to heal her legacy of mental ill health she did a ritual of giving back to her ancestors, as advised during an ancestral divination by Mbali Creazzo. She took herself on a long walk on the Pacific Coastal Trail at Lands End in the Bay area of San Francisco. 'I brought some offerings for them, some alcohol, and along the way I picked some wildflowers and stones to give to them. There is a labyrinth there and I walked it three times, all the time feeling that they were with me. In the centre I put all my offerings. Then I climbed down the cliff to the sea. It was pretty dangerous, as the tides were high. I took all my clothes off and got into the frigid water. I asked that the water send this energy back to them and I felt all of this pain and grief draining out of my body. It was definitely a turning point for me.'

The ancestors bear responsibility for what they have created and they are willing to take it on as long as we ask them to. Sometimes we find ourselves trying to fix our families. There is always one in any family who takes on trying to counsel, heal and help other family members.

Mary had always been the peacemaker in her family. Running backwards and forwards from one feuding faction to the other, she found that she was neglecting her own life. Finally, her family found itself at an impasse over an inheritance that favoured one family member over the others. The infighting had reduced her widowed mother to an emotional shell and Mary had run out of patience. In desperation she went to the graves of her mother's parents and grandparents in an overgrown cemetery in north London. It was the first time she had visited the grave since the death of her grandfather ten years previously. She took flowers and candles and sat by the graves for a while. Then she began talking to them. She told them what was going on in her family. She described the pain and loneliness of her mother, the anger of her sister and the dysfunction of her brother. Then she said, 'I no longer know what to do. I ask you to relieve me of this burden and to find a way to heal my family. I ask you to comfort my mother and bring her peace before she dies. In some ways it is you who have created this situation and so, from this moment on, I hand it back to you.'

Then she asked for a sign. 'It would be good to know you have heard me, so please give me a sign that I can understand that you are on the case.' She didn't need an immediate sign but suddenly she felt a presence to her right. She turned and saw a fox standing there, staring at her. They stood looking at each other for a few minutes in total stillness. If the fox had winked at her it wouldn't have been more obvious that this was an indication that her ancestors had, indeed, heard her prayers.

A few nights later she had a dream. She was in her mother's house, which had been beautifully redecorated, and the house was filled with children having a party. Her mother was in good spirits and was telling Mary and her two siblings (who hadn't been together at her home for five years at that point) about a celebration she was planning. Mary woke feeling that an era of joy and new life had begun for her family. 'Since the healing we have been brought together to love and accept each other just the way we are.'

Gaye Donaldson, a family therapist, confirms the liberation at the heart of giving back what is not ours to carry. 'What helps the ancestors the most is when we leave their fate with them. When we look and feel into the depths of what their own life journey cost them – the struggles, the losses, the hardships, the joys – and then, head bowed, we step back and respectfully leave all of it with them, then we dignify and honour their suffering. And we, in turn, take our rightful place as simply the recipients of life handed on, at exactly the price that it cost those who came before us. This is how we honour and respect the dead. If we wish to also please them, then we live a good and happy life and the ancestors see that nothing they endured was in vain.'

Forgiving your ancestors

It is time to forgive ourselves and forgive the others who came before us. Our ancestors seek our forgiveness to release them from the past and, by forgiving them, we free ourselves from the unhappiness we have inherited.

The more you are able to forgive, the more your life can be lived.

THE PRACTICE

- Place a candle on your family altar and pray to your God or Spirit for awareness, forgiveness and compassion.

- Call on your ancestors both ancient and known. Tell them about your hardships and pain; tell them how some of their actions have caused unhappiness in your family.

- Then say, 'I forgive you, I forgive you, I forgive you and I ask you now to help me heal myself and the rest of the family.' If you are unable to fully forgive, then tell them, 'I cannot forgive you at this time but I want you to know that this is my intention and I ask you to help me do this.'

- Sit in quiet meditation, focusing on your breath. Breathe deeply and relax as you breathe out.

- Close your eyes for a moment to feel what forgiveness means to you. Write down your thoughts and emotions. What are the changes you would like to happen for your living and deceased family?

- Allow your heart to be filled with grace and compassion. Focus on releasing the past by forgiving those whose actions have hurt the family. Focus on the candle flame and let the past go: what is not yours, let it go. Breathe deeply, into your heart, and release. Thank your ancestors.

Healing the Generational Legacy of Trauma

Perhaps it is only in subsequent generations that trauma can be witnessed and worked through, by those who were not there to live it but who received its effects, belatedly, through the narratives, actions and symptoms of the previous generation.

MARIANNE HIRSCH

How much did my father suffer from his experience of the Hungarian uprising, his exile and the traumas suffered by his father and two uncles killed in the war? He never talked about how his family suffered from the Soviet occupation but I know they had everything taken from them and were reduced to living a step away from poverty. My mother, too, is a descendant of the Spanish Civil War and was separated from her mother for four years after her father was executed.

Was it their shared legacy of loss and violence that brought them together in exile? And is that partly why I've chosen the work that I do: helping people to heal the past and rescue their ancestors?

Both my parents found it difficult to talk about the past. They were trying to create new lives in Britain and had little insight about how their life experience had been affected by their traumatic history. Both Hungary and Spain suffered under regimes that necessitated silence, so at home and abroad they locked away their memories and focused on survival.

The twentieth century was marked by a series of massive conflicts. From genocide in Armenia to the First and Second World Wars, the Vietnam and Korean Wars and further genocides in Cambodia and Rwanda, entire generations have been engulfed by traumatic events. How do we begin to heal the cycle of violence and trauma that repeats itself generation upon generation? How do we remember the dead, the wounded and the emotionally traumatised in our families, and how do we honour them without succumbing to bitterness, hate or depression?

One solution lies in being aware of the suffering of our ancestors, understanding their story and honouring their sacrifice, so we know who they were, why they died and how we, their descendants, can ensure they did not die in vain. And, in understanding their suffering, we can turn ourselves towards peace and reconciliation.

Historical trauma is transmitted from one generation to the next. Children of war veterans display symptoms of post-traumatic stress, even if their parents do not; the descendants of victims of genocide live in a constant state of anxiety even if their lives are comfortable and secure. In Germany, many of the children of the Nazis still live in an atmosphere of silence and denial, while descendants of the Armenian genocide are bombarded with the stories their parents and grandparents tell as a bulwark against forgetting.

Often it is the women who carry the loss, as they are the ones left to mourn the fathers, brothers and sons killed in war and genocide. Left alone with or without children in extreme situations, they sometimes choose to marry men they don't love as protection or are simply worn down by grief and hardship. As mothers, they were the ones who carried us in their wombs and nursed us in circumstances that were less than ideal.

The descendants of genocide spend their lives overshadowed by an event that took place long before they were even born. Much of the literature by children of the Holocaust expresses how the single most defining moment of their lives occurred before they were born: Auschwitz, Bergen-Belsen, the death camps and labour camps imprisoned them too.

Each situation is different, of course, and some descendants of these tragic events have gone on to lead successful lives, marry happily and bear healthy children, but for others the nightmares of the past refuse to let them go. The important thing is that people have an opportunity for collective mourning and remembrance, that they are able to tell their stories and that the loss is validated and honoured by the wider community. When this does not happen it is very difficult for the descendants to integrate what has happened to their families. When it does, we are more able to turn ourselves towards healing our family tree and the family tree of humanity

The wounds of war

So we may further our own healing and break the chain of trauma within our families, so we may create awareness of how a veteran's trauma ripples out over succeeding generations and end the silence about war's long-term human costs.

LEILA LEVINSON

The story of the veteran returning home, changed and sometimes unrecognisable as the one who left, is told countless times from countless wars across the centuries. The violence of war does not stay on the battlefield; it comes home with the soldiers.

Terry's uncle survived fighting the Japanese in Burma in the Second World War, but he returned home a different man. This working family man of gentle disposition had become aggressive, violent and spiteful. He regularly beat his son and his wife and he would go into psychotic episodes. The family just kept quiet and walked on eggshells around him, never knowing when he might spin into one of his rages.

Anthony Browne, the Children's Laureate in Britain, recalls the deep shadow his father's war experience cast over his childhood. 'I found his diaries after he died; he wrote about killing German guards with his bare hands – a shocking thing to discover about your gentle, loving father. My mum told me that she came into the house one day and found him wrestling on the floor with the vacuum cleaner. When he came to, he said he'd thought it was a German.' Jack Browne died of a sudden heart attack when Anthony was just seventeen.

Generations of soldiers have returned home with post-traumatic stress disorder (PTSD), undiagnosed and untreated, and we are only just beginning to discover how this affects their children and subsequent generations.[1]

Harmoni had been insecure and fearful most of her adult life and it had impacted her career as an actress and photographer. When she sought an ancestral divination, it became clear that she was suffering

---

1 Traumatic war experiences contribute to higher rates of cardiovascular disease, hypertension and gastrointestinal disorders. Rates of alcohol and drug abuse, domestic violence and mental illness are higher among veterans than any other group. In America, an estimated 260,000 veterans are homeless at some time of the year.

from trans-generational war trauma. When Harmoni's grandfather came home to America after the Second World War, he had become so angry and abusive that he even pulled a knife on her grandmother in front of his children. Harmoni's father was the youngest and felt his father's abuse most keenly. At just six years old, he learned early how to retreat inside himself.

Harmoni's parents divorced when she was two, so she only saw her father during the summers when she was growing up. 'As a child I idolised and protected him, only to feel that I wasn't important or even wanted. I know now that he finds it difficult to connect to people close to him.' The emotional disconnection from her father left Harmoni feeling as though nothing she ever did was good enough.

In the divination, Mbali Creazzo told her, 'Your grandfather has come to seek your forgiveness. He is sorry for what happened, he wants to reconnect and be brought back into the family. He is telling you that it is not your fault. You have inherited the violence of his abuse and the emotional disconnectedness of your father and that has impacted you with a lack of validation. Without validation, how can we know if what we are doing is right? This impacts our values of self-worth.'

Harmoni was advised to create an ancestral altar in her home and talk to her ancestors, saying, 'It's not my fault that things are the way they are and you need to take responsibility for that.' She was told to write a letter forgiving them and telling them what she needed to give back, and then burn the letter in a shell.

In the two years following her divination her relationship with her father has transformed. 'There has been an extreme shift, even a magical one, to the point that I now have the dad I always wanted. After the rituals there was an acceptance and forgiveness from me. I also believe strongly that the ancestors had a strong hand in his healing. I told my father that Grandpa had asked for forgiveness so he was aware and believed in the process as well.'

Annie can trace her family's alcoholism back to her great-grandfather's traumatic return from the First World War. This impacted her grandfather, whose trauma in the Second World War impacted her parents and finally her brother and herself. British politician Shirley Williams recalls the shadow the First World War cast over her mother, Vera Brittain, who lost her fiancé, her beloved brother, Edward, and many close friends in the trenches. She recalls how she doted on her elder brother, John Edward, who closely resembled her uncle in his artistic sensitivity and love of music. 'The legacy of war cast a permanent shadow over her life which nothing could quite dispel,' she writes. 'Four years of war had accentuated her natural trait of anxiety ... It was hard for her to laugh unconstrainedly; at the back of her mind, the row upon row of wooden crosses were planted too deeply.'

An estimated 10 million soldiers died in the conflict known as 'the war to end all wars'. More were killed on the battlefield than in any war before. Over 7 million men were missing in action, their bodies never found. Families mourned their beloved brothers, sons and fathers buried in foreign lands without the usual funeral rites, contributing to a surge in Spiritualism as many tried to make contact with those who had died. Many of those that did return home were disabled and traumatised. In those days, a deep shame accompanied what was then called shell shock and is now recognised as PTSD.

Latent war memories can continue through the family line and arise in unforeseen ways. The parents of a four-year-old girl brought her to see psychologist Anne Schutzenberger in desperation. Every night since her birth, she had suffered from severe asthma attacks and anxiety. Schutzenberger asked the girl to draw what was making her so afraid at night. She drew something that resembled a diving mask with an elephant's trunk. When the family went back into her family tree, they discovered an ancestor who had been gassed in the last gas attack on Ypres, on 26 April 1915. Coincidentally, or not, the girl had been born on 26 April. Once the event had been spoken about in the family, her nightmares and asthma attacks ended.

In decades of research, Schutzenberger has identified 'unconscious transgenerational repetitions' that often manifest around anniversaries. From 1992 to 1994 she experienced a surge of clients suffering symptoms that she believes were the effects of unfinished mourning in the descendants of those who died during the Second World War. The symptoms seemed to coincide with the fiftieth anniversary commemorations of war.

Today, wars in Iraq and Afghanistan are exposing another generation to the ravages of life-long trauma. Battlefield medical technology saves many who would have died in the past, but multi-deployments and IEDs have exacerbated PTSD symptoms. They call it 'the war after the war' and many are defeated by it. In 2010, for the second year in a row, more US troops committed suicide than were killed in combat in Iraq and Afghanistan combined.

In Mozambique, traditional methods of helping individuals and communities ravaged by the fifteen-year-long civil war have proved far more effective than talk therapy as healing tools. At battlefields across the country rituals were done to appease the spirits of the dead who had died far from home and without proper burial rites. These *mpfhukwa* spirits are thought to be disruptive, even dangerous, if they are not settled in the afterlife.

Crucially, the wounds of combatants were experienced collectively as part of one community, so everyone was involved in helping them let go of the trauma and re-enter civilian life. 'Trauma healing,' writes Alcinda Honwana, 'is perceived as a collective affliction affecting not only individuals but also their relatives, both living and dead. The objective of their cleansing ceremony is not to ignore past trauma, but to acknowledge it symbolically before firmly locking it away and facing the future.'

Dr Edward Tick, himself a victim of transgenerational trauma due to an upbringing with a war-wounded father, is a clinical psychotherapist who has begun to use traditional indigenous ways to help veterans in Britain and the US. 'I saw that war was so horrific, so traumatising, so damaging to every part of the body, mind, heart and spirit, that we needed extensive holistic-based practices for healing it. This wound is far bigger than we can cope with left to our human devices. We need spiritual connection, guidance and support in order to heal. I call my approach "spirituality in community" and it always includes memorials, always includes the recognition of the ancestors and other rituals to heal the past that we are all carrying and, if the dead still exist somewhere, they are still carrying as well.'

When vets come to him suffering from nightmares related to combat he takes a spiritual approach. 'I suggest to them that if they are seeing dead people in their dreams or nightmares, it may be that person's soul trying to contact them. I tell them that even if they don't believe it spiritually it doesn't matter, because it's true psychologically as well. If you are involved in taking a life or the loss of a comrade, you have an ongoing relationship with that soul whether they are here, in another dimension, or just inside you.'

When a medic came to him still suffering nightmares about the people he had been unable to save in an emergency ward in Vietnam, Dr Tick worked with this approach. 'We did two years of internal, imaginary dialogues with each person. The victims all blessed him and told him to live and to live for them as well. Some gave him tasks: one asked him to visit the Vietnam Memorial and say a prayer by his name. Another said, "You promised to visit my family and you never did. Do it now." Eventually we cleared out that emergency ward as he came to terms with each of his losses and the nightmares ceased.'

In many instances Dr Tick asks his patients to build an altar. 'I tell them to put photographs or mementoes of the fallen on the altar and use it as a meeting place, a place of focus. I ask them to tell the spirit that is bothering them to stop coming at night and instead come during the day during their altar time.'

The Plains Indians have a ritual at death called 'keeping and releasing the soul.' They build an altar to the dead and tend it for one year. Then they take it apart, burying and giving away the objects representing the person and saying prayers to release the soul to the universe. 'I do this with the vets as well, so they tend their altar to the dead who are haunting their dreams and, whenever they feel that they are at peace with this soul, we will take apart the altar and release them.'

Dr Tick has learnt much from his trips to Vietnam. 'Despite massive violent trauma throughout the country, there are minimal levels of PTSD. It is their spiritual practices and honouring of the ancestors that have enabled the Vietnamese to integrate massive injury and loss without breaking down.'

He takes vets back to Vietnam on sacred journeys of pilgrimage and healing. Bob Cagle served as an infantryman in 1965 and 1966 and kept dreaming of a young soldier he had killed. Then in a Buddhist temple on a mountain, Cagle looked up and saw the boy. 'I think we were talking to one another on some level I don't get,' he said. 'I felt like this kid could finally go wherever he was supposed to go. That's when I really started healing.'

The men fighting in Iraq and Afghanistan often have grandfathers who fought in the Second World War, in Korea or in Vietnam, and many of them have been unable to share their stories of what happened. Their descendants' sense of duty and familial honour, as well as their education and expectations, give them little choice but to sign up, but how prepared are they mentally and spiritually to cope with killing and death? The true cost of war is counted in the silent suffering of countless veterans and their legacy of hidden trauma and psychic wounding.

'In traditional cultures, warriors always came back to tell their stories and to do healing ceremonies in front of the entire community. The community witnessed the stories, felt the emotions, carried the burdens with their warriors and transferred responsibility for actions from the warriors to the community. In our society we leave our warriors the burden of suffering as if it were their own, when we should recognise that this is not their individual wound, it is ours as well.' Tick has created listening circles where veterans share stories with each other, their families and other civilians and so are reintegrated with the community.

Veterans of the Second World War are now in their eighties and nineties. There is still a brief and precious opportunity to listen to their stories, some of which they may be telling for the first time, before they go. As they divest themselves of their memories – some glorious, some burdensome – it lightens their load while making us all the richer.

The aftermath of civil war

The effects of civil war are measurably worse than conventional war, as the 'enemy' may be a member of your family or a schoolyard friend. Perpetrators and victims might have lived next door to each other. The muted celebrations that marked the 150th anniversary of the end of the American Civil War in 2011 reflected the bitterness still associated with it. Perhaps it's not surprising that the more recent civil war in Spain and the dictatorship that followed should remain so unresolved.

After the war, my grandmother was taunted in the streets and called a 'red bitch' or 'Republican whore'. As a Republican, she was prevented from working but she moved in with friends and her brother bought her a sewing machine so she could make some money. The women of Spain were left to bring up families alone, unable to grieve openly, committed to silence by the State. For two generations women lived in a constant state of anxiety and fear.

My grandmother knew one of the men who had killed her husband but she would never say who he was. She wanted to protect her daughters from hatred and somehow, despite everything, she kept her dignity and never descended to bitterness. She was a loving presence in all our lives, but she never spoke about what had happened. She kept it secret, like many families in Spain. For years nobody said a word. It was like it had never happened. Even today it is difficult for the older generation to talk about the past.

Consequently, there is still a deep sadness in Galicia. Whenever I return to Spain I feel this overwhelming emotion and find myself wanting to cry all the time. Even as a child, I would feel this without understanding why. Galicia suffered disproportionately from the post-war repression as Franco (supported by the Catholic Church) set about eradicating intellectuals, artists, teachers and Republicans who held political or legal power, like my grandfather. Men and women were taken from their homes and shot. Their bodies were left at the gateways of graveyards or on roadsides, in ditches and mass graves.

Over a million Spaniards were killed during the war and its aftermath. Unlike other countries that have tried to bring justice through truth commissions, Spain has never closed this traumatic chapter in its history: the atrocities committed by Generalissimo Francisco Franco's right-wing military squads have never been investigated or tried due to a blanket amnesty. And the dead still lie in unmarked graves.

José Molinos, a restaurateur in London, believes the shadow of the Franco years still lies over Spain. 'People want to find the bodies of their ancestors and give them a proper burial,' says José, 'but there are influential people determined to stop that happening. They have records, they know where the bodies are buried but they refuse to let that happen. Spain is still a divided country.'

So when the left-wing government passed the Law for the Recovery of Historical Memory in 2007, banning Francoist rallies and enabling the repatriation of exiled Republicans, among other measures aimed at reconciling the past, it was opposed by the right wing and the Catholic Church, who claimed it was 'opening old wounds'. And when the charismatic Judge Baltasa Garzon moved to open an investigation into the deaths of 113,000 Spaniards executed by Franco's men, he was suspended for exceeding his authority.

But the younger generation are refusing to forget even if their parents are too afraid to remember. Emilio Silva Parrera, whose grandfather was assassinated, founded the Association for the Recovery of Historical Memory (ARMH) to find, exhume and rebury those missing in action. The organisation has conducted forty excavations and found the remains of over 2,000 victims. His grandfather's remains were found in one of the first mass graves to be excavated.

Debate still rages over Franco's mausoleum, the Valley of the Fallen, which took more than eighteen years to build using some 15,000 Republican prisoners of war, many of whom died during construction. Under the monument is a mass grave where up to 40,000 unnamed Republicans are buried. Families have been lobbying the government to remove the remains of their loved ones or to move Franco's remains to a private cemetery and turn the monument into a national park of reconciliation and peace. 'That will never happen,' says José. 'It would cause another civil war. But something needs to be done. At the moment the victims of the right are celebrated and honoured. We need to honour the others and give them a decent burial. This is what Spain needs for real reconciliation.'

Phillippe, who lost his grandfather, grew up in the fear and silence of the aftermath and he feels it still. 'People suffered in silence for years but the old resentments simmer just below the surface.' As the namesake of his grandfather and his first grandson, Phillippe feels the weight of his legacy. 'For as long as I can remember, I had a sense a deep wrong had been perpetrated that I had to put right. I felt an anger that wasn't mine. It was only after having children that I knew I had to let go. But I still wonder if he is at peace.'

Since 1945 civil wars have caused the deaths of over 25 million people. Lingering shame and fear stunt the mourning for their loss. But the dead want to be remembered and they need to be honoured for us all to find peace. When I stood at my own grandfather's grave, I felt this was all he wanted. He wanted to know that his life had not been lost in vain. So many other grandfathers and grandmothers lie in unmarked graves throughout Spain and, until they are honoured by their country, Spain will be for ever haunted by their memory.

Healing from genocide

> From generation to generation we will declare their greatness.
>
> **AMIDAH PRAYER**

The children of Holocaust survivors – or the Second Generation as they are known – are particularly susceptible to the transmission of trauma, due to the extreme suffering of their parents. Either they were subjected to an unrelenting silence on the subject or they were deluged with a torrent of painful information about the past. Both are equally difficult to deal with. There is often a sense that they are atoning for the past, making up for it or healing it. They have to find ways to deal with their legacy before they are able to move on with their own lives. As well as feeling deep compassion and love for their parents, the Second Generation, often named for those lost to genocide, can feel resentment and rage in addition to guilt and inadequacy for not living up to their parents' expectations and not having suffered as their parents did.

Leila Levinson began her website Veterans' Children after realising her depression was linked to her father's. He had never spoken about his experiences as a GI medic who had been one of the first to enter the Nazi camps, but Leila discovered a stash of photographs after his death which showed he had treated victims of the most devastating of atrocities at the Mittelbau-Dora concentration camp outside Nordhausen in Germany. A relative told her that after treating the survivors, he suffered a nervous breakdown. 'I saw how my depression, my sadness, was a direct result of my father's. Despite his best intentions, he had transmitted his trauma to me. And my depression was passing to my children. It was time to break the chain.'

She decided that she would go on a journey to help her understand his journey. She approached other veterans who had liberated the camps. She discovered their silence had often been as impenetrable as her father's and their families had also suffered from these secret wounds. As they spoke to her, many for the first time, they wept as they remembered. The enormity of the suffering in the camps, the stench of death and the piles and piles of emaciated bodies, were images seared into their consciousness even when they were unable to share what they had seen.

Leila found the burial mounds her father had photographed when some 4,000 bodies were newly laid down. She describes how the reality of her father's experience hit her. 'After forty-eight years, grief pitched me off my feet. The difference between my father's trauma and my own dissolved.'

In her research, her writing, her pilgrimage, Leila reconciled herself with her father's legacy. She understood what had fractured his soul. Reconciliation means literally 'to meet again' and she met her father once more. 'My father walks alongside me now. He accompanies me as the sky accompanies me, as the ground holds me. I close my eyes and smell him, hear his voice saying my name, see his shoulders lift as he tucks the tips of his fingers behind his pants waistband, his eyes lighting up. He is not the father I knew when he lived. He has travelled with me, reclaiming his *neshamah t'hora*, his pure soul.'

Dodo's parents, who were both survivors of the Holocaust, each reacted differently. Her mother, whose father was killed in a concentration camp, talked about the war a lot, while her father, who lost both parents at Auschwitz, did not talk about it at all. Brought up in London, Dodo did not feel overshadowed by her parents' experiences but she recalls certain behavioural effects. 'Fitting in was extremely important and we kept our German side quiet. There was this sense of keeping our heads down. Yet we were very political because of my father. We were forever at CND marches and sticking up for the underdog. And there was no religion in our house, no belief in God. After all, our families had been persecuted because of their religion and, secondly, if there was a God, how could He have let the Holocaust happen?'

When her husband was working on the filming of *Schindler's List*, she took the opportunity to take her two daughters to the camp. While her mother approved, her father begged her not to go. 'I told him that I felt it was important they understood what happened.' It was the only time that they spoke about the Holocaust.

Dodo still does not believe in God but she connects with her parents who died months apart in 2005. 'I talk to them and I think I hear them talk back. The other day my daughter was delayed on her way from America to Greece. She was passing through Heathrow so I had a chat with them. "Okay, London is your territory. Help our baby make her flight." And the pilot held the flight for eight minutes so she made it.'

In many cases, Holocaust survivors found it much easier to answer questions from their grandchildren and consequently a close relationship has often arisen between the two generations. And now, as the Third Generation are becoming parents, they are making their parents into the grandparents they never had. The wheel of life continues to bring profound healing with each generation.

A huge ancestral branch of the Jewish race was removed in the Holocaust, along with millions of gypsies, ethnic Poles, homosexuals and the handicapped. The loss still haunts the world – and their descendants – and the remembrance of them in stories, films, anniversaries, memorials and museums is a vital way to maintain the links of the survivors to their lost relatives.

For this reason, it is even harder for descendants to handle genocide when its occurrence is unacknowledged. Actress, writer and director Sona Tatoyan has spent most of her adult life consumed by the Armenian genocide that devastated her family but is still officially denied by Turkey, whose Ottoman ancestors killed as many as 1.5 million Christian Armenians and sent at least 800,000 into the international community as refugees.[2] 'To have the genocide denied is to die twice' is a common sentiment among survivors, for whom the atrocities remain a painful daily reality, in part because telling the stories keeps memory alive. 'It is like a woman being raped but no one will acknowledge it,' she says. 'How do you begin to work through what happened if no one acknowledges that it happened? How do you get beyond the pain?'

The first genocide of the twentieth century began on the night of 24 April 1915 when the Ottoman government rounded up 250 Armenian intellectuals and community leaders and deported them. They were later executed en masse. Then the massacres began. Many were beheaded to save bullets, women were raped and their babies cut out of their stomach, whole populations were herded into churches and the buildings set alight. Finally, the remaining populations, usually women and children, were taken on forced death marches through the Syrian desert, where they died of hunger and thirst.

Sona's paternal grandfather was born in 1915 in the citadel town of Kharput, just as the atrocities were reaching their height. Her grandfather, an intellectual, was taken away and beheaded. Her grandmother, still pregnant or perhaps nursing Sona's newly born father, walked 300 miles to Syria. 'No one will tell me the story,' says Sona. 'There's so much shame associated with it. How did she make it? What did she have to do to survive? When I ask, they say, "They walked, they just walked."'

'My father remembers his mother would periodically put him on her shoulders and walk for miles out of the city, reliving the trauma that she endured.'

Sona uses her creative projects to transmute the stories of the past through her art. She recalls a recce to the Syrian desert with her husband, Oscar-winning writer José Rivera, and Micheline Aharonian

2 Figures vary, from 600,000 to 1.5 million people killed.

Marcom, author of a searing novel about the genocide, *Three Apples Fell From Heaven*. 'It was a surreal experience. There was a tiny shrine with a sign saying that a lot of Armenians had died there and that was it. Nothing but desert for miles and miles. We had heard there were piles of bones but we couldn't find them. Then a little Bedouin boy came out of nowhere and we asked him where they were. He reached down, picked up some dirt and in it there were these shards of white. "These are the bones," he told us. We were literally walking on the bones of our ancestors. At one point Micheline found the bone of an arm. It was one of the most appalling things. Watching this bone emerging from the earth. I had no words, no tears. It was beyond that.'

She is planning a documentary about these killing fields with a group of Turks and Armenians retracing the journey together, starting at the town where the marches ended and walking back to the heart of Armenia. It is as if Sona's ancestors are calling to her from beyond the grave, asking to be remembered, asking to be honoured for what they suffered.

'I feel that art can be healing and this is what I want to do with the documentary. I want there to be a catharsis for both sides and show the complexity of what happened. Not to say that the Turks did not butcher the Armenians but they were in a grip of an ideology, much like the Germans were with Nazism. I feel a lot of sadness for Turkey and try to imagine what it would be like if my ancestors were perpetrators and not victims. But they have to own what their ancestors did. If they don't, they will remain as stuck in this as we are.'

Research has shown that the continued denial of the Armenian genocide compounds the trauma suffered by subsequent generations, reinforcing feelings of insecurity, abandonment and betrayal. Denial is felt as a dishonouring of those who died and therefore blocks the grief and healing that need to be expressed and witnessed by the wider community. It also enables history to repeat itself. Famously, when Hitler was exploring his war plan to exterminate the Polish people in 1939 he said, 'Who, after all, speaks today of the annihilation of the Armenians?'

As with the Jewish Holocaust, an entire branch of Armenian ancestry was wiped out. 'Imagine,' says Sona, 'the great art that might have been created, the poems, the music, the books, the plays and paintings. It's an unfathomable loss.' In her projects she is giving them voice once more.

We need to find a way to honour our ancestors who died in these acts of brutal injustice without subconsciously repeating the patterns of history. When we fully acknowledge their sacrifice, we mitigate our subconscious loyalties that continue the cycle of violence and revenge. Dr Edward Tick suggests that descendants of victims of genocide make a journey of remembrance for their ancestors. By finding out their story, they begin to resurrect their memory and that heals their own lives. He sends them to archives and museums, to the Holocaust Museum in Washington DC, to Yad Vashem in Israel to seek out information about their ancestors who died. 'We have inherited incomplete stories that wound us and the more we understand what our ancestors went through, the more we will heal ourselves and help them heal. The more their stories remain tragic secrets, the more they harm us, but we can achieve understanding and forgiveness when we really comprehend what they have gone through.'

And, again, he suggests they create an altar. 'I suggest they make the altar quite public, put it on their mantelpiece in their living room so they can tell guests what it is and pause for a moment in remembrance. There's a healing power in making their story public. Our ancestors suffer less to the degree that we make their story known and useful to the living.'

But what happens when we are descended from the perpetrators of crimes against humanity instead of from the victims? Hilde Schramm was ten years old when her father, Albert Speer, began serving twenty years imprisonment in Spandau for Nazi war crimes. As a teenager she wrote a series of letters to him in prison, asking about his past and how he became involved in a regime of such darkness. She cultivated a relationship with him through these letters and also made a decision to atone for his crimes herself. 'We who survived the war are not guilty. We did not inherit the guilt, but the consequences of the wrongdoing of the past. We have to try and act with responsibility and one way to do that can be to try and give back.'

She sold his inheritance of three valuable paintings – which she believed had been bought from suffering Jewish families – and put the money in a foundation, Zurückgeben (meaning 'giving back'), to support Jewish women in the arts and sciences. Her motivation was simple. She said, 'It's about realising that the Holocaust casts its dark shadow over both time and generations and all the way into the nuclear family of today's Germany.'

She's not alone among descendants of Nazi perpetrators to face their family's past but she is rare. In some ways it is easier for the victims of war and genocide to express their grief and sadness; for descendants of perpetrators it is more difficult. How does one equate a loving father or grandfather with a mass killer? Many have chosen to remain in silence and denial, but there are those descendants who do not have any choice but to take on their family's legacy. The weight of history lies heavily on their shoulders. Their lives are dictated by what went before them. Often they are unable to fulfill their dreams, careers are stalled and relationships do not work. It is only by facing their legacy that they are able, finally, to find themselves and emerge from the shadow.

Healing the trauma of slavery

> I can still smell the spray of the sea they made me
>   cross.
> That night, I cannot remember it.
> Not even the ocean itself could remember.
> But I do not forget the first seagull I saw.
> The clouds above, like innocent witnesses.
> I have not forgotten my lost coast, nor my ancestral
>   language.
> They brought me here and here I have lived.
>
> NANCY MOREJON

The trauma of slavery lives on in the land and in the cellular memory of descendants who were stolen from Africa and suffered generations of horrific abuse. Somewhere between 10 and 20 million Africans were killed during the course of the slave trade.

Dr Booth Cohen tells the story of a young African-American woman who was travelling by train in Peru. 'It was a multi-hour journey and she passed the time looking at the landscape. Suddenly, she felt a strange disturbance in her body. Her emotions grew strong and dark, her heart rate increased and she felt overcome by feelings of stress and dread. Frightened and confused, she turned to her companion and asked what she was looking at. He glanced out and said, "Those are cotton fields." In recalling the story, she said, "Cotton is in my DNA. Even though I don't know what it looks like, somehow my body knows."'

How can we heal so strong a legacy?

There is nothing more powerful than walking where our ancestors walked, in the lands where they were born, lived, suffered and died. Their imprint still lies in the land and, if their legacy was traumatic, it can be immensely healing – for them and for us – to go back.

Myrna Clarice Munchus, a dancer and activist with African and American Indian ancestry, had always longed to go to Africa. But it wasn't until her grown daughter suggested she join the Interfaith Pilgrimage of the Middle Passage – a year's pilgrimage walking the route of the Atlantic slave trade back to Africa – that she knew it was time.

'It took us four and a half months to go from Massachusetts down the Eastern Sea Board to New Orleans. We stopped at cities, towns and villages, finding places where we could perform sacred ceremonies and offer songs, prayers and rituals for the spirits of those who had suffered and died there. We did this at auction blocks, underground railroads, lynching trees, whipping posts, plantations and slave quarters.' Along the way, Myrna was chosen to lead the prayers by pouring libations to the ancestors.

In Manhattan they were given access to a newly discovered graveyard where 400 slaves had been buried. 'We have a huge misconception that slavery didn't happen in the North but when we visited those burial grounds, it was absolutely stunning. We did our sacred ceremony, offered our prayers and invocations and then I lay on the ground and felt myself sinking further and further into the earth. The arms of my ancestors were pulling me closer. It was daylight but I felt I had gone into the underworld. I didn't want to leave. I just wanted to get closer to those who were calling me.'

When they arrived in Benin, on the West Coast of Africa, from where as many as 4 million slaves were shipped to the Americas, Myrna felt an immediate sense of homecoming: 'the sounds, the smells, the sights were all familiar to me.' Welcomed by the chief priest of Vodun, the indigenous religion, they were given a purification ceremony where the priest confirmed that many of them were the 'children of Vodun'. Myrna recalls, 'He answered my silent question as to whether my ancestors had worshipped in this way, even though my heart knew it was true.'

Her final gift was when he confirmed that she was the one 'who leads the people in prayer'. 'I was stunned, as this is what I'd been doing all along. All I had wanted to do was to go to Africa but it became an initiation and a spiritual awakening.'

Poet and writer Nef'fahtiti recalls a journey to Ghana, retracing the steps of her ancestors through the old slave dungeons, doing healing work with local priestesses. In the place where the most rebellious slaves were kept, surrounded by chains in the walls, they were told stories of what had happened. 'Suddenly, a rage came out of me that was so primitive, so deep, so all-consuming that I could hardly breathe. I could have killed with my bare hands. Quite clearly I remember hearing a voice saying, "You have a choice now. You can hold on to this rage or you can understand what happened." In a flash, I realised that to hurt, kill or maim another human being was a form of insanity. Then this compassion for them rose up in me. It was a big shift. Not that I don't have moments of outrage at racial injustice but it will never be as powerful. The ability to forgive really does change everything.'

Sid McNairy, a yogi in America, found his own way to heal the wounds in his family. Severely disciplined by his own father, he found himself giving the same punishment to his son aged four, beating him so hard that he couldn't sit down. 'That was the last time, as I made a decision not to repeat the same patterns.' He discovered that his father, a high-achieving academic, had been severely beaten by his grandfather who had been born a step away from slavery. 'I saw how destructive this anger was but also saw where it came from.' Today Sid's teenage son works at his yoga studio and they are 'not only father and son but best friends', and he has healed his relationship with his father.

There is a sweeter story in Sid's legacy that is part of his journey toward love and freedom: family legend tells of an ancestor, Thomas Hayslett, a white plantation owner, who fell in love with Missouri Lewelen, a beautiful Black Foot woman, at a slave market; he 'bought' her and her kin, married her and gave them all their freedom. This legacy led Sid to a propitious meeting with a Cherokee elder, Grandmother Morningstar, who helped him reintegrate his American Indian ancestry. 'Now I am proud of what is in me. My first pow-wow spoke to my soul. The faces, the skin, the love stood for itself. I was home.'

In America, 150 years after the civil war, the echoes of slavery remain. In a nation founded on racial inequality, racism still exists. However, the explosion of interest in genealogy, DNA testing and proliferation of research websites has meant that descendants linked by slavery have stumbled across one another and begun meaningful conversations around the past. In some cases white descendants have been able to hand over private documents – bills of sale, inventories, wills, letters – to African-Americans denied the thread of their lineage.

Susan Hutchinson grew up in the South, but was surprised to discover she was descended from slave owners. She found she was related to Thomas Jefferson, who was said to have fathered several children with his slave, Sally Hemings. In 1998, when DNA evidence all but proved this long-debated claim, Susan reached out to their descendants. She was honoured to be one of several white cousins invited to a reunion of the Hemings family held at Monticello, Jefferson's plantation estate in Virginia.

'It was a very powerful experience. After sharing our stories, we wrote a letter of apology and the next day, at a sunrise ceremony at the Monticello slave gravesite, we were able to express deep regret for our ancestors' participation in slavery. And express gratitude that we had been welcomed as family by his slaves' descendants.' They then held a ceremony at the Jefferson family graveyard.

Susan has co-founded a non-profit organisation, Coming to the Table – inspired by Martin Luther King's dream that the 'sons of former slaves and the sons of former slave owners will be able to sit down together at the table of brotherhood' – that enables 'linked' descendants to become a community working to transform the wounds of slavery. 'There is a great deal of healing for all of us to do. It is a brutal, terrifying history that tore communities apart.'

She discovered many slaveholder descendants who, like herself, had no idea about their slaveholding history because of the silence that follows each generation. 'It might be difficult to understand how slavery impacts the families of slaveholders but committing violent acts, whether sanctioned by society or not, also brings psychological and emotional trauma.' By telling their story, taking responsibility for the past and making amends, she hopes they can be part of a broader national redemption. When the descendants of Thomas Jefferson and Sally Hemings prayed and sang together, shared stories, hugged and cried, they showed that real fellowship is possible. Perhaps they are the beginning of a movement towards racial reconciliation in America.

Healing the trauma of colonisation

Native American isn't blood. It is what is in the heart. The love for the land, the respect for it, those who inhabit it, and the respect and acknowledgement of the spirits and elders. That is what it is to be Indian.

WHITE FEATHER, NAVAJO MEDICINE MAN

Communities around the world have been decimated by the colonisation of their lands and the suppression of their culture. Researchers have identified a phenomenon known as historical unresolved grief which contributes to the high rates of suicide, domestic abuse, violence and alcoholism among American Indians on the reservations. The same is true of most indigenous communities, where trauma is repeated generation after generation.

Recently there have been some attempts to stop the cycle of trauma by honouring the loss of indigenous land and human rights and reintroducing the spiritual practices that form the foundation for lives of harmony and balance. In 1999 some 100,000 hectares of land was returned to Khomani Bushmen of Southern Africa, arguably our oldest ancestors, in an unprecedented acknowledgement of their heritage. And in 2008, Kevin Rudd, then the Australian prime minister, made a heartfelt apology to the Aboriginal people for their mistreatment by successive governments.

In America, where American Indian spiritual practices were illegal until 1978, there is a resurgence in the use of the sweat lodge, peyote ceremonies, prayer gatherings and the sun dance. A new generation is finding the way back to the 'good red road' of the ancestors. In communities around the world it is a return to the ancient spiritual practices of the ancestors that is healing families and communities.

Mona Guitterez, a spiritual activist in Tempe, Arizona, recalls that it was her first sweat lodge at twenty-two years old that helped her stop drinking. 'It was my very first native ceremony and this combination of fire and water put together inside this womb, this earth belly, is so powerful. Truly I was reborn. I don't think I would be anywhere without the sweat lodge.'

Mona was invited to take part in a peyote ceremony which would be another pathway for healing. 'I had only been sober for three months, and as soon as I sat down, I felt as though I had come home.' Since then, she has inspired her family to return to traditional pathways of healing and empowered her community by facilitating peyote ceremonies herself.

The spineless blue-green peyote cactus has been a sacrament since the earliest days of humanity. Today, the Native American Church uses peyote, often in all-night teepee ceremonies, with fire ritual, prayers and songs. Peyote, a mild hallucinogen, is used by at least 250,000 people across fifty different tribes in the US today. 'In both the sweat lodge and in a peyote ceremony, we are healing the separation that was brought to the Americas with colonisation. During those ceremonies you know that we are all the same, that we are all connected.'

Mona welcomes the resurgence of these ancient indigenous practices in communities across the Americas. 'We are celebrating being alive, not just through our bloodlines, but through our culture: our dance is alive, our languages are alive, our songs are alive. And that is what is making the healing happen. Without these rituals and ceremonies, our people would be long gone. If we didn't have the sweat lodge, the drum, the songs, the dances, we would not have survived.'

The ancestors, they say, are returning and bringing back the wisdom and the knowledge that was forced underground for so many years. And according to Mona, they are also calling for the dissolution of the separation between races.

'As we move forward we will understand that being native has nothing to do with blood. We are all medicine to each other. Human beings are medicine: we come into the world as medicine for our parents; we are medicine to our brothers and our sisters and everyone we connect with. Being native is not defined by race, colour or nationality. Being native is being someone who cares for

Mother Earth. Being native is understanding that we are caretakers of the earth and we are going to need all the caretakers we can find.'

Zena Duze, a filmmaker, activist and healer, has spent her whole life seeking to heal the psychological and spiritual damage she suffered growing up as a 'coloured' in apartheid South Africa. 'We just pass it on and relive the trauma from generation to generation. I knew I'd been damaged but I wasn't aware of how deeply generational it was. In my family, the pain of trauma was covered up with alcoholism and so I was suffering both from the original trauma and from the alcoholism that resulted. It all contributes to the feeling that I don't belong or I'm not good enough. The shame and self-hatred go so deep.' She lost contact with her black father at an early age and, in common with most coloured families, she tended to identify more with the white side of her heritage and knew virtually nothing about her Xhosa or Zulu roots.

Her spiritual journey began with her calling to be a sangoma or traditional healer, where arduous training connects adepts directly with their ancestors. One of the most powerful moments of her personal reconciliation was a journey to Robben Island, the notorious prison island where Nelson Mandela was held for eighteen years. She connected with her ancestor, Chief Maqomo, a Xhosa freedom fighter, who died there in the nineteenth century. Known for his eloquence and intellect as well as political and martial skills, he was imprisoned for twelve years but, on release, he attempted to resettle on his land and was banished to the island again. He died mysteriously in prison in 1873.

'I could feel him there,' she says. 'I could feel his longing for home and for his wives and children. I could feel his anger at the dispossession of his people and his frustration. One day his spirit led me to this old abandoned building on the island that had once housed insane women. Then he said to me, "Only when [all] the women come will the healing begin."'

Now living in America, her grandfather's vision has encouraged her to work with women from different ethnic backgrounds. While she still cultivates a dream of building a women's healing centre on Robben Island, she creates weekend retreats for women and children: gatherings of story-telling, ritual, dance, art and great food. Usually filming and researching as she goes, she is finding ways to heal the generational wounds that racial politics have inflicted on communities around the world.

Journalist and anti-apartheid activist Max du Preez has written about the need for all South Africans to honour and recognise their ancestors, whatever their colour or ethnicity. His research into the history of his country revealed a king called Moshesh known for his compassion, wisdom and diplomacy. Du Preez calls him 'the Nelson Mandela of the nineteenth century'. His discovery of Moshesh made him aware of the importance of ancestors and the value of understanding those who came before him. 'I started to realise that I could not make sense of my society if I did not understand what events and which people shaped and influenced our attitudes and memories.'

An ethnic Afrikaaner, who was never comfortable with the world of apartheid his Dutch ancestors had created, he came to understand that he was above all an African and could lay claim to the great African leaders and visionaries of the past alongside the canny Afrikaaner guerilla fighter, Christiaan de Wet, who fought the British. 'You don't have to be in the same ethnic bloodline of the great men and women of the past to own them as heroes,' he says. He claims them alongside the more traditional Afrikaaner heroes like Piet Retief and Paul Kruger. 'All these people are my ancestors, the fathers and mothers of my nation who came before. I don't have to like them or agree with what they did for them to be my ancestors, they just are.'

He often exhorts his white readers to understand that they too can claim the Khoi leader, Autsomato and look to Shaka Zulu, Dingane and Moshoeshoe as their own ancestral heroes. And blacks can claim the great white activists Joe Slovo and Helen Suzman and others who were part of the struggle to create a democratic nation with one of the most enlightened constitutions in the world.

Like former president Thabo Mbeki in his magnificent inauguration speech, Du Preez asks that all South Africans honour who they are by honouring all the ancestors who contributed to the diverse ethnic mix of modern South Africa. 'We should claim our common history as South Africans,' he says, 'otherwise we will remain victims of the past.'

Noel's story

Noel Tovey, award-winning dancer, actor, choreographer and the artistic director for the welcoming ceremony of the Sydney 2000 Olympics, has spent most of his life coming to terms with his Aboriginal heritage. 'In the 1950s Aboriginals weren't considered human. My parents didn't want to be seen; they hid away, disappeared into alcohol and drugs. It wasn't until 1967 that we became citizens of our own country. "Little black bastard" were probably the first words I remember,'[3] he says. Now he mentors Aboriginal youth, teaching them about a heritage that goes back 100,000 years. 'I tell them about our knowledge of healing plants and the stories of the Tingari

---

3 Noel Tovey wrote his memoir *Little Black Bastard* in 2004. It is now an award-winning one-man play.

who come from the Dreamtime and taught us how to be human. I teach them to honour their elders. Our young people are hungry for this knowledge and I am seeing healing happening all around me. Passing the wisdom on is part of being an Aboriginal. We have to do it before we die and I know my ancestors are keeping me alive for this purpose.'

He heard his ancestors for the first time as a teenager. He was in jail, in deep despair and considering suicide. 'I felt something in the cell grab me,' he says. 'I saw images of my ancestors coming out of the wall and I heard their voices yelling at me, "Did you ask to be born in the slums of Carlton? Did you ask to be born black? Did you ask to be raped and abused from the age of four? NO!" the voices said. I didn't know what was happening to me. The voices got louder and my head started spinning and I passed out. I knew that what they had said was true. None of this shit was my fault. They were telling me to hold on, that there was a better life for me somewhere out there and one day I would find it.' He had rediscovered his indigenous soul and from then on his life began to turn around.

In 1960, he was able to afford his passage to England, where he had a distinguished career for thirty years, all the while denying his heritage until a friend accused him of 'fighting for every black cause except my own'. In 1990 he decided to return to his homeland and begin the search for who he was. 'Up until that point I never belonged anywhere but slowly I began to get a sense of myself.'

He discovered that his grandfather and great-uncle were in fact African American and had been famous minstrels in the late 1800s. They toured England and played for royalty, while his grandfather gave banjo lessons to the Prince of Wales, later Edward VII. They were the first black musicians to have their music recorded. His grandfather is buried in Brompton Cemetery in London where Noel used to go when he had a script to study. 'The music was in me from the beginning.'

In 2008 he had a profound catharsis when Prime Minister Kevin Rudd publicly apologised for 200 years of mistreatment of the Aboriginal people. He was speaking for generations of his own ancestors in a long, heartfelt speech that told the stories of Aboriginal children ripped from loving communities and delivered into institutional care. Some 50,000 children were brutalised as a 'product of the deliberate, calculated policies of the state'.

Kevin Rudd apologised for all of it and called for a reconciliation that would transform 'the way in which the nation thinks about itself, whereby the injustice administered to the stolen generations in the name of these, our parliaments, causes all of us to reappraise, at the deepest level of our beliefs, the real possibility of reconciliation writ large: reconciliation across all Indigenous Australia; reconciliation across the entire history of the often bloody encounter between those who emerged from the Dreamtime a thousand generations ago and those who, like me, came across the seas only yesterday; reconciliation which opens up whole new possibilities for the future.'

Noel, watching from a primary school, broke down. 'I couldn't stop sobbing,' he says. 'I was inconsolable. It became very personal. I heard, "I'm sorry you were abandoned, I'm sorry you were raped, I'm sorry you were jailed and mistreated." Then when someone asked a young boy how he felt about the speech and he said, "Now I can be like Uncle Noel and be proud to be an Aboriginal." So I started crying again. He was a great man for doing this. It was the beginning of a healing journey for our country.'

SIX
Our Inheritance

When a child is born it carries the light of all the ancestors of the past all the way back to the Creator. Every child born wants to know who they are so they can understand where they are going. Their past, present and future is intertwined and if they know where they come from it gives them the road to walk on in life.

GRANDMOTHER PAULINE TANGORIA

The story of the next generation begins at conception and continues as the child is birthed and enters the world: these are our first steps towards our own ancestral continuum. We are the descendants of our ancestral lineage. We are the miraculous manifestation of two lineages coming together. Aspects from each side of the family combine to create a unique individual and the beginning of another generation.

The marriage ceremony is a moment when we can understand ourselves as part of the ancestral continuum in a potent way. It brings together our living relatives, but the ancestors of the bride and the groom also attend and sanction the wedding with their blessings. After all, these are the future parents who will provide the progeny of our ancestors and continue the family line.

We inherit the ancestors of our partner. They too become a part of our lineage, whether or not we have children, as those ancestors will bless the partnership and make their influence felt. Carla Esteves, a doula and family therapist, believes that we come together to bring healing from one lineage to another. She says there are no coincidences in life but synchronicities. 'We come to earth as blood family and when our paths cross we become (soul) family. These encounters are meeting points offering a gateway for us to experience ourselves on earth more deeply, they are gateways offering the remembrance of oneness.'

Conceptions and birth

> When we welcome our babies with wonderment for their journey, love and respect for their individuality, and trust their knowing, we give them the true gift of life – to be themselves in the safety of their parents' love.
>
> **KITTY HAGENBACH**

The unspoken dreams and hopes of our ancestors swirl around us in the womb. Our genetic history and some of the inherited behaviour patterns lie dormant in the unconscious, ready to be reanimated, when we are standing at the doorway of incarnation. Although many of us will arrive into a stable, loving partnership, some of us come into a family under more complicated circumstances. We may have been unwanted, unexpected or followed miscarriages or abortions;

we might even have been the product of rape or an extra-martial affair.

How do we sense that we are the result of this type of conception? If the circumstances of our arrival were less than ideal, we might begin to look for clues to our parents' situation and emotions at the time of our conception. This atmosphere would have surrounded us from the start: in the womb we encounter our father's attitude to our arrival and his feelings for our mother; and we take on the influence of our parents' relationship and the memories of our grandparents and great-grandparents, the shadows of our ancestral continuum.

This time in our mother's womb begins to teach us about our family and their circumstances and prepares us for the place we find ourselves in once we arrive. Whatever was happening in your parents' lives at the time of your conception – historically, emotionally and materially – may be the source of much of your contentment or despair. If your parents were delighted by the news of your arrival, you are likely to have been a contented child; if you were unwanted or your parents were unhappy, this will also be reflected in your personality.

Like people in many indigenous communities, I believe that a particular ancestor takes spiritual responsibility for each child at the moment of birth. Conception, birth and death are gateways to the spirit world and the ancestors stand at these openings, like spirit guardians. Throughout each conception and pregnancy I had vivid dreams about certain deceased members of our family. The day my daughter Sequoia was born my husband's aunt died, and it was her father Henry, Terry's English grandfather, who came into my dreams. With my son Ossian it was my Hungarian grandfather Josef, and with Bede, my youngest son, it was my Spanish maternal grandmother, Beba. Curiously, each child shows characteristic similarities to that particular part of the family tree; this in turn has a powerful influence on their behaviour, physical and psychological tendencies.

Sjanie, a London-based hypnotherapist, recalls the support she received from her ancestors while giving birth. 'I saw all of the women in my family line completely surrounding me in a circle. It was so clear and so strong and so present and it felt like they were there to bring this baby into the world. My ancestors are South African, half English and half Afrikaans/Dutch, so they are all pioneers. I have always seen them as being these incredibly strong and courageous people and that was the feeling I had in my labour – of the overwhelming sense of this unfailing strength. I knew I was being supported by the strength that came from them, it never came into my mind to stop or give up.' Sjanie was moved by the magnificence of her maternal ancestry when she looked into the eyes of her daughter for the first time in that special moment at birth. A long line of maternal influences touched her in that special moment of birth.

In previous generations, dozens of women died in childbirth. More recently, mothers were often drugged and their babies whisked away to the glaring lights and sterility of the hospital nursery with no fathers in attendance. These generational birth traumas can be held in our subconscious and will reappear when a woman becomes pregnant, during pregnancy, at the birth or post-natally. Strange desires and fears will materialise that relate to an event or trauma that affected their mother, grandmother or even great-grandmother.

When Carla fell pregnant for the first time, her body started reacting as though it was in trauma. 'I was invaded by enormous waves of fear that no rational mind could explain. My whole body would shiver and shake and I started crying waterfalls. It shocked me that someone like me who feels herself so close to nature was so scared at the thought of being pregnant, the most natural thing in life.'

She realised that at least part of it was the fear of being abandoned in motherhood. 'My grandfather left my grandmother when she was pregnant with my mother. And then my father also abandoned my mother when she fell pregnant with me. I came to know that fear did not belong to me but had been passed down to me by my mother and grandmother.'

She then discovered that her great-grandmother, Rosa de Jesus, had fallen into a fire when she was nursing her newborn. Although she did not drop her baby into the fire, it did not survive this traumatic event.

'I came to understand how we repeat the stories of our ancestors until they are released, how these stories tend to be stored in the womb, and can be released at great initiations in a woman's lifetime such as birth, menstruation, pregnancy and menopause. I felt her pain in my own body when I was pregnant, the fear and agony that she experienced in those months.'

We can inherit the trauma of past generations without knowing it but once we become aware of what has happened, it can be released. Kitty, a psychotherapist, teaches parents how to bond with their babies. She believes that if there has been a 'secret' in the family you cannot resolve it in the present: you have to go back into the past and resolve it by bringing the memory out in the open and forgiving what has happened. When Carla experienced her great-grandmother's grief in her pregnancy she recognised that she needed to honour the life of her great-grandmother, and the trauma no longer had to remain inside her. In this way, the secret does not remain forgotten but is present today, in ourselves. Kitty believes that 'Once we make sense of our lives, and remember the lives of our ancestors, we can heal.'

Our actual birth may be our first lesson in whether the world is a safe and warm place to be. Each birth is individual and no two births in one family are the same. Sometimes a child just needs to be born in a more complicated way; it may be the ancestral agreement with their family, and this can trigger a personality trait that will determine them and their future. As we incarnate into our family, the ancestral shadows from our parents wrap around us like skins that protect, challenge and educate us through our early years.

As Laura Hayward, a doula from London says, 'It is the child's journey as much as the parents'.' Watching a baby in the first few minutes of life is fascinating: they all look so old and many very wise! You can see the spirit settling into the physical form, pulling in its aura to the confinement of the baby's body.

Mother love

> Yesterday I found a photo
> of you at seventeen,
> holding a horse and smiling,
> not yet my mother ...
> although I was clearly already your child.
>
> **OWEN SHEERS**

Our primary relationship is with our mothers, so their imprint on us is indelible. The relationship is rarely simple, as the emotional implications of being a 'good' or 'bad' mother are so very strong. There are selfless mothers and narcissistic mothers, smother mothers and absent mothers, controlling mothers and chaotic ones. Whatever form it takes, our relationship with our mother affects how much we value ourselves. Our parents are usually the most important people in our lives when we are young, so taking a close look at our relationship with them will teach us things we can be really proud of, as well as a few things that remind us what we are not happy with.

For women, our mother is our first role model and we will inevitably mould ourselves to her image even when the example is an unhealthy one. For sons, the mother is their first relationship with the opposite sex and can affect all their future relationships with women. How we negotiate our way through this experience depends largely on how our mothers and our grandmothers negotiated the labyrinth of motherhood. No matter how many modern books we might read on mothering, it is almost impossible not to fall into the template our mothers gave us. When a mother holds her child in her arms, whether a son or daughter, she feels many emotions. If that child is the same sex as she is, she sees herself. We feel every wave of every emotion that we have ever felt, from love to hate.

Georgia Coleridge recognises the challenge of becoming a mother. 'Before I had children, I had ambitions to be a wonderful mother. I'd be Julie Andrews on top of the mountain, smiling and playing the guitar, with everyone skipping along in perfect harmony. I would never get tired or lose my temper. My children would never be cross or upset. Of course the reality was totally different. I loved my four children passionately, but I was disappointed to find that I was often exhausted, furious and out of my depth. I wanted them to have a different childhood from my own, but it was a struggle to overturn old family patterns.

'My own mother was very highly strung with lots of problems of her own. She'd been barely brought up at all by my grandmother, who was a glamorous, absent alcoholic with a string of husbands. My great-grandmother was sensitive and artistic but frequently depressed. I came from a long line of well-meaning but emotionally clueless women. I was desperate not to pass on this baggage to my children.

'Luckily I had a lovely father, supportive husband and lots of determination. I read dozens of books, enrolled in parent classes for years and quizzed every happy mother I knew. I also did lots of work to heal my own faults and insecurities. I'm still not a perfect mother by any means, but I am no longer overwhelmed. Our family feels robust, and the relationships are now relaxed and easy. My children will have their own struggles in life, but I feel that the blueprint of that ancestral burden has altered forever.

'It is this deep knowledge of ourselves formed in childhood that has a huge influence on how well we cope at being a mother. Our self-esteem is of critical importance and is a primary influence on how we relate to our offspring. However we have been mothered ourselves, some good and some bad ideas of mothering will be passed down to us and then down to our children. We can celebrate the best of our mothers and heal the rest by being aware that we can change what we have been given as children and the cure in all cases is LOVE, deep unconditional loving Love.'

As we look back into our own family line we understand the influences that affect our choices and know that we can break that legacy. We inherit and pass on our genetic memories through the female line. Our mother forms her eggs whilst in the womb of her mother, we form our eggs in her womb and our daughter in ours. This can also mean that your grandmother's experience of giving birth can be reflected in your mother's experience. So if your mother's mother was anxious or afraid, these emotions might have impacted in utero. If you are a war baby born during the Blitz or during any stressful cultural or social upheavals, you would have felt it. If your parents were fighting, if there was fear of scarcity or dis-placement or if your mother simply had personal concerns about her own value, security or place in the world, all this would have left a powerful impression on your subconscious.

When we become parents ourselves, it is as if we awaken these memories with symptoms of strange phobias, desires and fears.

Fears of death augment the hysteria we bring to childbirth as another aspect of our collective unconscious, but these memories lie dormant until awoken by the physical experience of labour and the early days of motherhood.

Sara's story

Sara Bran, a journalist and writer, discovered that her own pregnancies and births were a reflection of what had happened to her great-grandmother. 'I have had two traumatic birth experiences. Both my daughters were born by emergency c-section after I had laboured fruitlessly for three days with my first daughter, twelve hours with my second. My babies were cut out of me while I lay there, utterly terrified and disengaged from the process of becoming a mother.

'After my father died suddenly just before my second daughter turned two and two weeks before my fortieth birthday, I plunged into what I can only describe as a black chasm. I realise in retrospect that I was entering a time of great healing. I explored all the stories and family myths I had inherited surrounding motherhood and childbirth. Along the way, I uncovered stillborns birthed alone on open prairies, children born in secret and surrendered for adoption, women whose hearts corroded at the kitchen sink as they entered motherhood without joy. I began to understand how ideas of birth can be inherited energetically just as tangibly as the colour of an eye or an aptitude for music.

'The story that really resonated with me came from my aunt who visited from the US after my father's death. She told me about my great-grandmother Irene, an Irish immigrant to New York in the 1880s.

Irene was a classic Irish redhead who loved her whiskey a little too much and was prone to fiery rages. She also suffered a horrific tragedy as she was birthing my grandmother. Her son came running into their home literally on fire as she was labouring. He had suffered some kind of accident and died as my grandmother was entering the world. Irene's subsequent depression led her to reject my grandmother, and she did not feed or nurture her – the consequences of which are so obvious to me now generations on.

'What echoed with me was the sense I had in labour that birth meant death and death meant birth; that the two were intrinsically linked and that if I let go, something terrible would happen. It is as if somewhere in my body I still held the horror even though it wasn't actually mine. I do still wonder what my birthing experiences would have been like if I had engaged with my ancestors earlier. Perhaps if I had heard their voices and honoured their experiences, I could have on a deep level understood that their story did not have to be my own.'

Our relationships with our mothers and their relationships with their mothers can sometimes trigger post-natal depression, unexplained fertility problems and sexual issues. Sonia, a counsellor and holistic healer specialising in women's health, has spent most of her adult life dealing with the legacy of growing up with a troubled and abusive mother. Her mother, suffering from unresolved anger and low self-esteem, veered from spoiling her daughter to beating her severely. Sonia was desperate to become acceptable to her mother but this drive for perfection resulted in ulcers and insomnia, and later she developed anorexia and bulimia, both diseases of control and self-reproach. When she met her future husband they moved from

South Africa to the UK, putting some much-needed distance between herself and her mother.

It wasn't until she and her husband decided to conceive that the old issues resurfaced. 'One moment I went from feeling completely ready for motherhood to suddenly having really big fears. Every fear, insecurity and negative belief that I had carried about myself from my childhood came back. I cried more than I can remember in my whole life; I felt more anger than I knew I was capable of.' She was barely able to work, eat or focus on anything and her libido all but disappeared.

After several traumatic long-distance calls to her mother she realised that all the women in her mother's line of her family were damaged in some way and Sonia acknowledged that she had to resolve her issues before she had her own child. 'Something must have happened somewhere along the line to initiate their deep self-loathing, and repulsion for their femininity and sex that continued to be carried down the line of women. Was it abuse? Maybe. I don't feel I really need to know. All I need for now is to know that this is transcended and healed. And for me and my children, new beliefs around femininity, sexuality and self-love can be laid down.'

Clinical studies have shown that any negative beliefs that a mother has about herself, her spouse and her circumstances during pregnancy can all have an effect on her hormones, confidence and bonding with the pregnancy. Any family history of early deaths of mother and babies, sickness or abuse can amalgamate in the subconscious, creating anxieties and fears that seem abnormal. Post-natal depression is a term that can be used to categorise this disorder. This loss of confidence and fear can permeate your whole being. And although in most cases the child will appear not to be affected by their mother's depression, the effects can lie dormant until later life, until an age when they might consider having children themselves.

Sometimes, those who observe or absorb their mother's inability to cope might decide not to become parents at all, and will make a clear decision not to have children. Janine made a decision not to have her own children because she wanted to stop the dysfunctions from continuing forward. 'It has taken me a while to understand what we carry and what we have to let go of before we can flourish – I carried at least three generations of abuse and chaos and I knew at an early age that it would all stop with me.'

Kitty's story

Kitty is a psychotherapist who works as a pre- and peri-natal therapist. She realised that her early desire not to have children was based on the fact that she was not wanted by her parents.

'My father had stated that he did not want children and would take no responsibility if my mother got pregnant. He kept his word and openly rejected myself and my brother. I grew up in Ireland and despite marrying young as everyone did at that time, I was very clear, I did not want children. So why was I so determined not to have children?

'I thought no more of it until years later after I had left my first marriage and began a little introspection. I found myself attending a business conference. I was sitting in the back row of a lecture theatre, and I heard a loud clear voice say, "What you really want are children." I turned around to see who had spoken these surprising words to me and of course there was no one there.

'Could this be true? I came to the understanding that I had made up my mind not to have children while I was in my mother's womb. I was shocked to think that I could take this decision even before I was born. So what was the problem, what was the basis of my fear?

'I discovered that neither of my parents was wanted, and both were only children. My maternal grandmother was sent away when she was born. She returned when she was adult and accomplished. My grandmother in turn left my mother when she was born and went on a six-month cruise to recover from the birth. My great-grandmother had eloped, so safe to say there was probably not a great maternal infant bond there. I traced my maternal line back seven generations, discovered there had been poor to non-existent maternal infant bonding all the way down the line. I now understand that we mother as we have been mothered unless we take steps to become conscious of these inherited patterns.

'As I was pondering if I would be able to become a mother, I had another magical experience. I suddenly had a flash of seeing myself with a baby who was about six months old. The Universe was holding my hand so caringly and helping me to overcome my deepest fears. After this my second husband and I decided to open up to having a child. I conceived easily, against all the predictions for my age. It was thrilling to give birth to our first son, I was completely happy. We were blessed with a second son two years later, both of us love being parents and becoming a mum is the very best thing I have ever done in my life.'

Just as infertility or the decision not to have children can be rooted in inherited patterns, trauma such as the loss of a mother during childhood can cause women to struggle with becoming a mother themselves. This happened to Holly: 'I lost my mother when I was sixteen years old. It had a profound effect on me at the time and it goes on to shape every decision I make as an adult, albeit subconsciously. The thought that she would miss me having children registered at the time, but I had no idea then just how profound her loss would be to me when I did go on to have my girls. Or indeed how much I would continue to feel her loss as I struggled to become a decent mother myself.

'I am learning to accept myself as their mother and I am learning that motherhood can be wonderful. I want to be the best mother I can be, I want to be whole, not diminished by my past depression and grief, as I believe if I let that define me it will ultimately define my children. My sadness that my own mother never saw my two beautiful girls is deep. But it is balanced by the love that these two special people are bringing and will go on to bring moments of intense joy to my life. It feels as if this is my new beginning, my time to have a life shaped by happiness and fulfillment, not sadness and loss.'

Sometimes issues with our mothers remain, as we become adults and mothers ourselves. Agatha's relationship with her mother has always been challenging, so she devised a ritual for cutting the ties with her mother, as she wanted to heal their relationship. 'During the ten days of conducting this ritual, my mother, who did not know anything of what I was doing, was ringing me from France, every day, asking me if I was okay. At the end of the ten days I had this dream that my mother and I were both stillborn babies kept in a bottle of formaldehyde. I saw myself in the dream breaking the bottle and releasing the babies. It symbolised that we had emotionally separated, that I no longer needed to carry her legacy, which I naturally did as her eldest child (I am one of seven children). I always felt that I carried my mother's ancestral shadow: all the morbidity, heaviness and suppressed emotions from my mother's family, who are wealthy French aristocrats.

'After conducting the ritual, I found it easier being in her company, we became close and very supportive of each other. I definitely had more patience and time for her. It changed the way I felt and in response she became much kinder, more gentle with me.'

Cutting ties, to release and heal

Many of us have suffered from difficult, mainly challenging relationships with someone from our childhood. Often the person is our mother or father, but can also be a sibling or grandparent.

This exercise will help you come to terms with difficult family relationships, either with a living relative or a deceased family member. It is particularly useful when you have negative feelings about someone or there has been a long-term estrangement that has not been expressed or healed.

Before you begin write down the answers to these questions:

Who do you want to forgive?
What has hurt you, when and why?
When did it begin?

PREPARATION

- Light a candle on your ancestral altar and ask that the wise and compassionate ancestors help you let go of your past wounds and hurts and open the heart of the person who has hurt you, so that they too can let go.

- Write down what you wish to say to the family member who has hurt or angered you. As you are doing this, let any thoughts or feelings of anger or hurt find their way on to the page. When you have finished, do not read what you have written.

- Take the piece of paper, fold it up and burn it in the flame of a candle or on an open fire. Let it go into the flames to burn away symbolically. Say to yourself, 'I am now free of this hurt and pain. I release it to the ancestors for their help to heal past struggles in my relationship with (name).'

You are now ready to forgive and let go of the past. The next step is the ten-day ritual of cutting ties.

THE PRACTICE

Conduct this exercise every night for ten nights, sitting quietly for half an hour before going to bed.

- Sit and breathe; focus on peace.

- Draw a big figure of eight on a piece of paper and put this on your ancestral altar, with a photograph of you in one loop and one of the person you want to cut ties with in the other. Place a lit candle in the middle; leave the candle burning while you meditate.

- Look at the photograph for a while, then close your eyes and become mindful.

- After you have done this, write a summary of your thoughts or the feelings that you hold inside you. Let any feelings and thoughts go and, with your breath, focus on seeing yourself releasing those ties.

- Ask yourself how you can change your relationship now. How do you want to be able to feel about them?

- Observe the rise and fall of your emotions as each day passes. Be aware that your emotions may move from forgiveness and love to anger and rage and back to forgiveness and love: these cycles of emotion are the natural progression which a healing or cutting ties meditation can generate.

- At the end of the ten days place the photographs together, light a candle and place it next to a small vase of flowers to bring love, healing and peace.

- Leave the photographs together for twenty-four hours, asking the ancestors to guide and protect this relationship. Go to your ancestral altar, light a candle and give thanks to your ancestors who have assisted you in cutting ties.

HEALING INHERITED FAMILY ISSUES AT YOUR CONCEPTION

As you explore what was happening in your family, you may realise that there was some dysfunction at the time of your conception and during your time in the womb. If you feel that your parents were unhappy or you were unwanted, this simple meditation can help heal and forgive the past.

THE PRACTICE

- Find a photograph of yourself as a baby and place it beside a photograph of your parents or parent. Place them on your ancestral altar and light a candle.

- Pray for the memory, the emotions of your conception to be blessed, and forgiven. Ask your ancestral guardian to place a seed of light inside you, to release and heal any dysfunction or negative memory at your conception.

- Leave the photographs on your altar for seven days, lighting a candle in the morning and evening. Write a letter to your parents telling them how you felt as a child – remember what was good about your childhood and what was bad. Tell them that you forgive them.

- On the seventh day either go to the place that is sacred to you and bury the letter or burn it. See the past draining away into the earth or burn with fire. Say to yourself, 'I am now free of this inherited pain and heartbreak. I release it to the ancestors for it is now theirs to deal with, so I can move on and be at peace.'

Father love

> I saw behind me those who had gone, and before me
> those who are to come. I looked back and saw my
> father, and his father, and all our fathers, and in front to
> see my son, and his son, and the sons upon sons
> beyond. And their eyes were my eyes.
>
> RICHARD LLEWELLYN

The father of the family is a powerful archetype: whether he is present or absent, he looms larger than life in our imaginations and our hearts. It takes a long time to see a father as anything other than hero or villain. In some primal way we know that we can never really lose our mothers, so we can strike out easier against them when we are angry or upset; but we know that our fathers can more easily walk away. They become mythic beings, casting their shadow over our whole life. For boys, the father will teach them how to be men; for girls, this primary relationship will predicate who she chooses in a partner.

The role of fathers is changing rapidly in western society as fathers take on more responsibility in the home, share child-rearing duties and spend more time with their families. But like women, they are still bound by old legacies. They may find that their sons trigger latent insecurities and cause them to react in anger or be less supportive than they would like. They may not recognise that the shadow of their own father is awakened at the moment they become a father themselves, and that it will be a reflection of the way they were parented. This is the main challenge for fathers. Generations of men never spoke about their relationship with their children and many had no relationship at all. But there are many good men who instinctively know how to be a father by tapping into their ancestral pool of good fathers.

When Drew became a father for the first time in his late twenties, he immediately felt the deep ancestral memory that connected him with his lineage. 'I attended the birth of my son, Malakai, and actually caught him as he came into the world. Fatherhood kicked in at that moment. Suddenly I was really focused and completely tuned in and there was a tremendous tribal feeling from deep inside me.

'My greatest realisation in becoming a father is that as a couple we had created the next generation of my family. Estelle is an only child, so is my mother and so am I and so is Malakai. But I did have that sense of being related to a tribe, as my grandmother was one of thirteen. I had always wanted siblings but at least now being a father I created a new generation for our tribe.'

My husband Terry believes that his experience of becoming a father was very different from that of his own father. 'My father was an archetypal man from his generation, who had just come out of the Second World War. He had served in the Royal Navy and although he loved being at sea, he never spoke about what he experienced during the war. He then went into the fire service working shifts and from time to time taking on a second job just to pay the bills. He was pretty absent during my childhood. So when I became a father I had to learn how to do it differently, I was working from home and a hands-on dad with all our three children.

'My father never attended my birth, yet I was at the birth of all our children but the most profound for me was when our daughter was born because she was our first. You could say I was startled when I saw her, overwhelmed and I wasn't prepared. When she was born I felt that time momentarily stood still. I was in awe of the birth, a miracle of a child radiating this feeling of "I know I am welcomed and loved". That was exactly how I felt too, as if we both knew a long time ago that we would arrange this meeting, so for a few years only she would be in my care. I am proud of all my children and what they are becoming.'

Nicola's husband Rob came to parenthood from a different perspective. The father of two boys, he believes his joy in parenting is a direct gift from his own father. 'I can trace a lot of parenting gifts to my father: his warmth, his encouragement. But the thing I marvel at most was the freedom I felt in fighting with him. We had shouting matches more times than I can remember – but I never felt intimidated by my father. I think that translated into a pretty secure adulthood for his son and grandsons – who are now young men in their early twenties. I see where I got my confidence and I see where my sons get theirs but I've been mystified by where my dad got the confidence to trust his children, let them go their own way – sometimes against his wishes – and instinctively know how important that was.

'My father had lost his father at about age nine. The few memories he shared were of a troubled man filled with worry and rage. But I guess somewhere along the way my dad got the message that fatherhood was a joy, not a burden. Where did that come from? I don't think it's something you can easily learn on your own. I believe it's something you just know. And somehow he knew it.'

Something like the trust that Rob describes could be the product of many different ancestral traits. His grandparents were refugees from the pogroms in Eastern Europe. Their experience taught them to stay close to their families in case they had to flee for their lives at a moment's notice. His great-grandparents may have enjoyed the security of life in a small scholarly village where faith in God blanketed the entire community in a belief that everything would turn out all right.

Sometimes a trait such as trust doesn't descend in a pure form from one generation to the next but is made up of a hodgepodge of seemingly unrelated traits from the near and distant past. The more you can understand your ancestral evolution, the better you can understand the character traits that have shaped you and either put them into practice or change them to benefit the next generation.

Statistics on 'fatherless' children – the ones born of fathers who are absent through divorce, death, alcoholism or drugs or who are emotionally remote – show that they are more likely to be expelled from school, take drugs, have mental health issues and commit suicide. They are less likely to be gainfully employed, fulfilled or successful. The loss of the father seems to result in children who have been cast adrift, unable to make their way in society as easily as their peers who have been well fathered.

A father sometimes inherits his own father's erroneous ways and passes them on. The lesson for all fathers is to recognise their infants' individuality and try to lovingly support it. This is most difficult if a son is the extreme opposite of this father, as not only will the father feel a failure in his guidance but will also wonder how opposites can ever understand each other. In such cases, both father and son have to work courageously on the examples they set for each other as the child matures, because ultimately a child of any age still needs his father's love and approval, while the father seeks to be proud of his son whatever the differences that separate them as men.

The way to heal a wounded relationship is to forgive and accept each other the way we are. If we can forgive a dysfunctional and absent father we can heal so much of the past. Compassion and forgiveness can prevent us repeating the same cycles of dysfunction, as we let go of our unfulfilled expectations about our father when we were children. If we accept the fact that in some cases they cannot be our hero, we can then stop classing them as our villain. We can forgive them, repair what has happened and liberate ourselves.

For Mark, the fact that he was adopted overshadowed his childhood, but becoming a father himself has been really healing for him and he is now reconciled with his biological father. 'My adopted and biological fathers were great fathers but I have always been aware of being adopted, the insecurity and sense of never quite fitting in. Becoming a father has healed this for me and I realised the importance of fathers: that they are there to make us feel safe. Our father's love is a validation of who we are and that gives us the confidence to forge our way in the world, make mistakes and pick ourselves up and carry on.'

Normally a child who has lost their father has no outlet for their grief and for the terrible pain and social difficulties that losing a parent can bring. When Nicola's father died she says it was as if someone had ripped away the foundation beneath her. 'My father was the rock of our family and as he disappeared from our lives it would be that wound that would forever define me.'

Karl's father left his family when he was nine years old and he has resented his father ever since; he felt abandoned and wounded by the rejection, particularly as his father moved countries and set up a new life with another family. 'You never recover from the pain of rejection and as a man I realise that I had no ideal male archetype to follow, as my mother never remarried and never forgave him for leaving us. It was when I settled down and became a father myself that I knew how he must have felt leaving us all behind and I needed to heal this between us. We met as adult men, talked about what happened and why he left and didn't return, and I found a way to forgive him and changed from hate to love.'

In some cases there is no reconciliation and the wounds go from generation to generation, with many boys being brought up in single-parent families with challenging masculine role models. Their father's physical and emotional absence makes it very difficult to learn to be good fathers themselves, but there are ways to break their father's legacy and not repeat history. One way is to look further back into the family tree and find a grandfather who is a more supportive role model.

Sarah's story

Sarah's partner Stephen was from Guyana, and Sarah is from Somerset in England. Sarah and Stephen have very different cultural and ethnic roots. When he died suddenly from a congenital heart defect, aged thirty-five, he left Sarah four months pregnant with their daughter Stevie. Sarah wanted to ensure that her daughter knew her father's heritage. She said, 'I want her to have a real sense of who her father was and keep the connection with his bloodline.'

Her mother Mary told me that what happened to Sarah had also happened to her mother and grandmother. 'I lost my father when I was four years old and now my youngest daughter Sarah lost her partner. My grandmother was left a widow, then my mother became a widow and was left with three young children to look after. I felt too young to remember my father, but my mother never spoke about it and just got on with her life. I am grateful I have never been widowed myself but the added tragedy of Sarah carrying an unborn child instilled a grief so powerful that I thought it would drown us. It was a nightmare. History, horribly, had just repeated itself.'

Five months later Sarah gave birth to a beautiful, adorable and feisty daughter who symbolised new life and hope. She was young and beautiful, with her whole life ahead of her. And just like my mother, who had been very stoic, was not over emotional and got on with her life.'

Sarah made sure that she saw Stephen's family. Sarah told me, 'I have never pushed the memory of Stephen away. I have photos of him around the flat and I talk to him all the time as well as talking to Stevie about her dad.' Like her grandmother Mary, Stevie will not have any memories of her father but she will heal her loss by knowing that her family heritage and her connection with his family will always support her. Sarah says, 'Stevie's character is so strongly like Stephen's siblings. She is from a powerful matriarchal family. They all have similar characteristics and I know that when she grows up she will always know who are father was and embrace that to be able to pass that on to her own children.'

Adoption: non-biological heritage

For many people, adoption is a controversial and emotional topic. There is still an assumption that the mother-child relationship is primarily one of biology. This is not always true. In many cases, adoptive children identify most strongly with the adoptive parents. Maybe it is the element of being 'chosen' that makes the link between parents and their adopted children more spiritual.

Heidi already had a grown-up child when she decided to adopt with her new husband. She discovered that the path to adoption was as challenging and demanding as being a biological mother. 'There are many signposts along the road to adoption. It is clearly a journey, not a destination.' After two heartbreaking failed adoption attempts – she calls them 'miscarried adoptions' – she found that her adoption journey mirrored the pregnancy process in many ways, even down to dreaming her connections with her future children. When it looked as though the courts might stop her third attempt to adopt two Russian girls, her future children came to her in a dream. 'I saw both Liza and Lana and asked what they were doing. They very calmly said to me, "We are preparing to come home with you. You are our mother." I woke up – startled. Could this be true? When I saw Liza, she recognised me and immediately said "Liza Mommy" and Lana pushed herself against me as if her very life depended upon it. As the court date approached, the dream came back to me over and over again. I knew that they were coming home.' And they did. 'I know of no separation between birth and adoption, between my dreams and reality. My reality is my dreams and being a mother is the greatest honour on earth.'

Julia was adopted in Ireland at a time when villagers would crowd outside the adopted child's new home and pray for the child's soul because the 'bastard' had been 'conceived in sin'. She also suffered considerable stigma at her private school but Julia has never doubted that she belonged to her adopted family. 'Just before my mother's death we had one particular emotional exchange that I will always treasure. Whilst in hospital unable to move because of her medical condition, she was deteriorating rapidly, unable to speak very clearly. She was determined to tell me how much she'd always loved me and how she hoped she had given me a happy life. I miss her a great deal – if that's not a "real mother" then I don't know what is.'

Julia points out that many people spend a lifetime attempting to come to terms with the unsuccessful relationship they have had with their biological parents. 'My elder sister, the product of my mother's first marriage who was adopted by my dad, has always had an intensely difficult and volatile relationship with my mum. It was in every way opposite to mine. Most of us adoptees know that we have been chosen, so it is as much an integral part of me as having green eyes.'

Sometimes simply finding the biological family, knowing who they are and understanding their story is sufficient to make peace with the past. My sister-in-law Natalie recently found out that she had a half-sister. Her mother Eileen went on a website in her recent search for her adopted daughter and put down all the family details. Natalie said, 'Within twenty-four hours I got an enquiry from a Sally. She had been looking for us for over twenty years. My mum was sixteen years old when Sally was born in 1960.'

My brother Paul remembers taking them to a hotel for their first meeting. 'Sally is exactly like Natalie and physically similar to Eileen. They spoke with the same voice intonations, hand gestures; their presence was so uncannily alike.' They have all remained in touch and they have a close relationship, which Natalie believes is what Sally was always looking for to feel that she belongs in our family.

Adopted children often seek out their birth parents at times of transition, such as when they become parents themselves or when one or both adopted parents dies. Those who have been adopted know well that their own bloodline influences their personality and physical mannerisms. Adopted children are profoundly aware that they are not the same as their non-biological parents and siblings.

There is no doubt that our genetic heritage affects us regardless of whether or not we have been brought up by our biological family. Adoptees who traced their biological roots as adults often discover an uncanny number of similarities between themselves and their far-reaching ancestors. Adopted at birth, they could not have been consciously influenced in their choice of a particular profession or in the attributes and traits that they share with their bloodline, yet this phenomenon of ancestral similarity can be found again and again. Although knowing their family is out there and just like them can bring comfort, some discover their biological family and then run back into the arms of their adopted parents: they feel spiritually that their soul is more linked with them.

Kate had a very good relationship with her adopted mother. 'She died a few years ago. I really loved her and was very loved by her family. Unfortunately I cannot have children. I have unexplained fertility problems which may or may not stem from my adoption and a deep-rooted fear of rejection. Since the death of my mother, I have been on a spiritual journey and discovered that my adopted ancestors – all the females in my family line – pay close attention to my life and I feel them around me when I need help or support. I have their photographs on my family altar and pray to them when I need help. I don't really feel my biological ancestors with me, as physically I have had no connection with them that I felt was supportive or caring.'

Of course, the opposite also happens and in Mark's case he feels more akin to his biological father than his adopted family. 'My adopted parents and all my siblings are very, very different to me. Nature and nurture are completely obvious. I now have a very good relationship with my biological father but I realise that without the support of my adopted family I wouldn't be as successful as I am today.'

'My adopted parents raised me, so a lot of the values I carry came from them but the inherent backbone of who I am is my biological family. You cannot change the DNA inheritance and I think that is a huge frustration. I find it incredibly easy to get on with Patrick, my biological father, and his entire family, but with my adopted family I still find it bloody hard to get on with them as we are so different in lifestyles and attitudes.'

Adopted children are very aware that they are given away but family is family and love is love and their ancestral connection will come to them from the family that gives them the greatest support and care.

Grandparents: honouring our elders

Grandparenting is a valuable time for the bonding of
love and understanding between the generations.

MARY VANDERHOOK, A GREAT-GRANDMOTHER
AND GRANDMOTHER OF 13

The old are the precious gem in the centre of the
household.

CHINESE PROVERB

For most traditional cultures an elder is acknowledged as having
reached not only a state of old age but also a state of maturity and
wisdom. The elder is as important to the family and community as
a newborn, and they share the proximity of just coming from and
going to the spirit world. In traditional society the elders of the
family or tribe are held in the greatest respect as with the graduation
from parent to grandparent comes the honour of being closer to the
ancestors, and so having more wisdom and compassion.

Becoming a grandparent is a graduation to that special place in
the family tree, no longer having the responsibilities of bringing up
a family yet enjoying the fruits of their children's labour. There are
good and bad grandparents, absent ones, caring ones and inspiring
ones – yet most are idolised by their grandchildren and accepted
with all their idiosyncrasies. We love our grandparents because
they are there to spoil us, love us and accept us for the way we are.
Grandparents in general have no expectations and we are perfect in
their eyes from the moment we are born.

Romana's grandmother was from Zimbabwe and was a powerful
and deeply intuitive woman. 'When I was growing up I loved her
and was scared of her – she had this way of looking at you as if she

could read your thoughts, she would know when something was going to happen or if any of us, as in my siblings or cousins, were in trouble.'

Our memories of our grandparents are potent as they are essential to connect us with our heritage. Grandparents teach us the things our parents don't have time for – such as little games or sayings from a previous era. They pass on information about ways of life that may have died out but were once part of our culture or folklore. They may have been the only source of unconditional love we received when we were growing up. As a child I would go and spend my summer holidays with my grandmother. I felt a deep connection and unconditional love from her. I believe that she is still a powerful benevolent force of love, for all the family.

Danny had a very close relationship with his grandmother. 'I am one of five grandchildren and she was my father's mother, born in 1914. She was very strict but also incredibly naughty, and out of all the grandchildren I was her favourite.

'When I was twenty-one years old, I decided to travel around the world. Before I left it was just impossible to say goodbye to my Nan! I kept coming back into her flat to say goodbye and was crying my eyes out. At the time I wondered if I was going to see her again.'

Often there is a powerful spiritual link between grandparents and grandchildren: when something happens to either of them they will know it subconsciously, no matter how far apart they are, and Danny knew when his grandmother passed away. 'We were living in Australia, six months into the trip and we used to call home once a week. One night my sister and I were in a bar and I suddenly had a really horrible feeling – I knew something had happened to Nan. I said I had to leave and call home straight away. Dad answered the call and said my Nan has just died. He later told me that she had wanted to die in my bed at my parents' house. I was completely devastated, wondering if she had died because I wasn't there for her and she was lonely. She will always be a huge inspiration for me.'

By talking to those who are still alive we can hear their stories and then pass them on to our own children. Stories are far more valuable than the most expensive heirloom, as they tell us who we are and help us and our children find their way in the world. There is an African saying that when an elder dies it is as if a whole library has burned to the ground. Talk to your grandparents and great-grandparents if they are still alive. Listen to their stories. Today we're seeing the Second World War generation reach their eighties and nineties. As we have seen, many did not talk about their lives when they returned to civilian life but now they may be ready to tell their stories. Record them, and you will have preserved something that will be treasured in your family for generations.

As parents we learn how we naturally inherit the memories of our parents and grandparents, including their rules, values and traditions, and how we love both unconditionally and conditionally. Freeing ourselves from our parental conditioning begins when we stop denying that we are like them and instead start asking ourselves how we feel, act and react in the same way as them. This is the beginning of choosing with awareness what we pass on to our children and grandchildren.

Our cultural heritage

> All human beings today are the products of the co-evolution of a set of genes – almost identical across cultures – and a set of cultural elements which is diverse across culture but still constrained by the capacities and the predispositions of the human mind.
>
> **JONATHAN HAIDT**

Our relationship with our cultural background may not be apparent when we look at immediate family. Nowadays, many of us are hybrids, born in one place but with one or other parent from somewhere else. We are affected by both the heritage of our parents and the dominant culture in which we grow up.

In Britain, there are now at least two generations of British-born Afro-Caribbeans from the West Indies. Similarly, there are Indian, Pakistani, Bangladeshi and Ugandan children who may never have visited the country of their grandparents. Some speak only English – others are bi- or tri-lingual. They might listen to British music and eat British food but there remains a strong pull toward their own ethnic background as well. Rituals, beliefs, prejudices, strengths and the idiosyncrasies that shape how you live may be rooted deep in the past.

We are talking about generations of habits and rituals being handed down. Baz is British-born first generation from Cypriot Armenian parents. 'Our parents insist every member of our family speaks Armenian at home and will correct each other's pronunciations to ensure the perfect memory of the language. I think my parents are afraid that we will lose our culture if we didn't work at remembering.'

Just as genes transmit biological information, so memes carry cultural influences. In modern western society, as people move away from their family home and migrate to other parts of the world, away from their culture, these memes could change for our children's generation unless we encourage that connection. Genes have a physical basis, but a meme takes the form of an idea, behaviour or style that spreads from person to person within a family and within a culture. Memes convey our inherited behaviour and belief systems about religion, language, money and food. These beliefs are transmitted through what you read, how you speak, gestures, rituals and behaviour. If there is no consistent encouragement of particular belief systems, they can become extinct or change, depending on the circumstances of the following generation.

The beliefs that we inherit about ourselves and our lives are not always true, because they belong to previous generations. For instance, in my family there is an integral belief that it is essential to keep to a healthy diet and know how to cook from the freshest and most natural ingredients. I now recognise that a fear of getting sick underlies this sensible thinking. I have discovered that this obsession with health and food reaches back to my great-grandmother, who inherited polycystic kidney disease. Many of her children and grandchildren also died from it. For them, a good diet and strong health were extremely important for their survival.

This method of inheritance can also occur in families where there are eating disorders, sexual abuse and fertility issues, money fears and anxiety. For instance, someone may believe the 'cookie theory' and think, 'I have an eating disorder because my mother punished me for stealing from the cookie jar. She punished me because her grandfather punished her.' This story probably began with someone in the 1500s that had a cookie issue. Once we see that the story is a story and not a fact, we can begin to distinguish the difference between what has been told to us and what is really true: we can begin to disentangle ourselves from our family's script.

Food and cooking

Immigrant families may change the way they dress, the way they worship, the music they listen to, but they bring their recipes with them, teach their children how to cook and often change the tastes of their new mother country.

I have always lived in Britain, but it is my Galician heritage that rules my taste buds and my cooking habits. The peppers of Padron, a speciality cooked in the best olive oil and sea salt, is still my favourite dish. They are mainly sweet but there is the occasional hot one that leaves the rest of the family shrieking with laughter. And I notice that my son has taken up our family habit of sucking on prawn heads (which horrifies my English friends).

Mexican girls in America are still brought up to make the perfect tamale, Indian women are still judged on the perfection of their rotis. For Jewish families the most sacred family moments are almost always linked to food. Even in fast-food America, ethnic food is lovingly prepared in millions of kitchens. In a country where TV dinners and pizza prevail, families celebrate their most important holiday, Thanksgiving, sitting together around tables groaning with home-cooked food.

Romana remembers the ritual hierarchy of cooking in her Muslim family in Zimbabwe. At seven, she was given the chutneys and pickles to make, then she graduated to cooking the rice and finally perfected her special dish of curried chicken. 'Cooking is considered an art and girls are expected to be able to cook from an early age. Once you find your specialty you are then expected to cook it at family gatherings.'

Food is our comfort, our taste of home. Cooking and eating brings families together and it is around the table that we experience the highs and lows in family relationships. Our connection to food roots us to our heritage and is one of the most tangible traditions that we pass on to our children.

Religion

The majority of our ancestors within recorded history were brought up in religious communities that revolved around churches, chapels, synagogues, mosques and temples. Before mass migration, the vast majority of western society considered itself to be Christian. Religion shaped the way ordinary people lived, provided the moral framework and defined the attitude of communities. It also had strong influences over sexual behaviour, contraception, illegitimacy and women's rights. Religious festivals provided the majority of the holidays and Sunday was a day of rest across the country. My family is Catholic and this deep sense of God and ritual was a part of my childhood. I still adore hearing church bells ring, as it reminds me of been taken to church by my grandmothers and hearing them say their prayers in their native tongues.

Not everyone remembers their religious roots in such a positive way. Darius, an Iranian lawyer living in Los Angeles, grew up despising Islam. His family were intellectuals connected to royalty and would have nothing to do with the Ayatollah. They left after the revolution in 1979. Darius married a fellow Tehranian who had no such problems with her Islamic roots, taking from it what she wanted and leaving the rest. She used her religion as an expression of her natural spirituality and love of ritual. So she prayed and honoured the fasting of Ramadan and the rituals of Narooz. Their differing attitudes were purely a matter for light-hearted (mostly) dining-table debates until she gave birth to their first son and the question of circumcision came up: he took against it as a form of anachronistic mutilation but for her it was a matter of cultural and spiritual continuity. She wouldn't send her son into the world without the ancient rite of passage that would set him apart from all his cousins, uncles, grandfathers and great-great grandfathers. In the end, their son was circumcised.

Not long afterwards, in a surprising twist, Darius was sent a book about his ancestors. In it were pictures of generation after generations of imams. 'There was this row of brothers wearing turbans, all ayatollahs with these huge beards. They all looked like the Taliban to me.' Something must have happened in the past to destroy his family's strong link with Islam. It is interesting, as there is a passion in Darius's anti-religious zeal. He is, in fact, quite religious about his atheism.

Nicola's family isn't religious but every night when her father was home in time he would come and kneel with by her bed and say three prayers. Always the same three prayers, probably as his Irish father had done with him and his father before him. Even though she would not call herself Christian, she thinks those early moments communicating with an invisible being called God set early spiritual pathways that have enabled her to continue having a relationship with the invisible.

Religion and spirituality are some of the most powerful belief systems that are passed down in families – even, as in Darius's case, when the idea conveyed is that there isn't a God. And we still gather around religious festivals like Christmas, Easter, the feasts of Yom Kippur, Eid and Diwali.

Money and prosperity

Our ideas of prosperity, of abundance and scarcity are deeply rooted in our family's attitudes about money. And they are difficult to shake. Was your family a 'money grows on trees' kind of family? Or a 'you have to work hard for everything you get' family? Or was it the kind of family that assumed abundance was a birthright? This kind of thinking can shape our entire lives if we let it.

Morgan was brought up in a Calvinist family where she was taught that you had to work hard for every penny. She's a writer, organised and madly talented, but has constantly found herself unable to access the prosperity that her work deserves. She finds herself overworking for far less pay than she deserves while the fear of 'not enough' has lodged itself deep in her subconscious, so that she goes for jobs unworthy of her talent just to earn a living. This then destroys her self-esteem and creativity and continues the cycle of 'I am not worthy' or 'I am not enough'.

Livia, by contrast, believes that money does indeed grow on trees. Her background gives her a confidence that when she needs it, the money will arrive. And usually it does. Her belief in prosperity seems to materialise financial support.

Past situations in our family often contribute to our complex relationship with money. Georgina's father can never hold on to his money: he is always losing it or giving it away. The characteristic is so marked that Georgina suggested they find out if there was any related family history. He comes from a long line of vicars and they discovered one who had profited personally from the children's charity set up by his parish. The profound guilt had passed subconsciously to Georgina's father. In Jennifer's case, her father had been born into a mining family in Wales and could not shake off the poverty of his childhood, although he had become a very wealthy self-made man. 'We had all this money and this huge house,' she says, 'and yet there was always a feeling of scarcity.'

If we look back over several generations we might witness a steady accumulation of wealth or vast fluctuations. Certainly in America the memory of the Great Depression hangs over families who went from prosperity to destitution overnight when the stock market crashed on 29 October 1929. Breadwinners jumped to an early death from skyscrapers and bread queues stretched for city blocks. Economists have noted that during the first decade of the twenty-first century, the disparity between rich and poor grew to its greatest since the 1920s, just before the Great Depression. History ignored is history repeated.

Inherited wealth creates its own issues. 'Shirtsleeves to shirt-sleeves in three generations' refers to the tendency for one generation to lose what a previous one has gained. So where one ancestor makes a huge amount of money, his son might spend it all in a life of profligacy and return his son to the position his grandfather was in before he made the money. Only forty-eight years after the death of Cornelius Vanderbilt, the wealth builder of the family, one of his descendants died penniless as succeeding generations demonstrated an ability to spend money with breathtaking recklessness.

Warren Buffett, one of the richest men in the world, has famously decided not to leave his children a significant proportion of his wealth. He is opposed to the transfer of great fortunes from one generation to the next and once commented, 'I want to give my kids just enough so that they would feel that they could do anything, but not so much that they would feel like doing nothing.'

The family home

Our home is the key to our security and our family's roots. We have strong memories of the places where we lived when we were growing up. Often we take these memories into adulthood and into our own home and our own family life. Did your family remain in the same home or did you constantly move around? Did you stay in the same area as your parents and grandparents lived, or did your family migrate far from its roots? This all has an effect on the way we create our own family home and pass on the legacy to our descendants.

All properties hold memories of past lives, and sometimes negative memories – death, loss, financial ruin – come with the building. But when we bought our house in Somerset, we knew straight away that most of the families who had lived there were content and happy, and our previous house in Scotland also had a lovely feel to it.

One day we had a knock at the door and an elderly couple with their daughter and her husband asked if they could come in. They were visiting from Canada but were originally from Scotland, and our property was where they had given birth to their daughter and brought up their young children. As they walked through the house they stopped and peered into the downstairs bedroom where Ossian had been born just a few weeks earlier. The mother told me that she had given birth to her daughter in the same room fifty years ago. We were amazed at the way life had come full circle. Our families were strangers yet briefly felt a unity, a continuum.

Stately homes, manor houses and old farmhouses where many generations of the same family have been born, lived and died, carry centuries of emotion and political and religious intrigue. Emma's family had owned their heritage home for generations, but few of them had been happy in the house. Emma discovered the house had been built at the expense of another's misfortune and also that there had been a murder in the house. However, when her brother Tom, as head of the family, finally decided to hand it over to the National Trust, Emma was initially devastated. She was filled with rage and grief as she felt her ancestral legacy disintegrating. But during a healing session her grandmother came through to tell Emma to let it go. Emma remembers, 'She said it would be best for everyone and that it would liberate us and then she asked, "What are you so upset about? It's only a house!"'

The house has been turned into a museum and since then Emma and her siblings have felt a sense of liberation. 'My brother has since married, had a son and lives happily in Devon, my sister too is doing very well, and I am back to painting and creating my art. So my grandmother was right – now the house is no longer ours we are really truly free.' She has since gone back and made her peace.

By contrast, actress Greta Scacchi always felt rootless. Her peripatetic childhood split between Italy, England and Australia left her feeling fragmented so she was determined to give her children a sense of permanence. 'I thought it was a curse not to have one place that was home so I wanted my children to have their own valley, their place, their home. I wanted them to have their childhood memories rooted in one place so they would grow up feeling secure and have a long safety cord that they could use to do whatever they wanted to do and go wherever they wanted to go. Whether they will appreciate it or not, I don't know.'

It is interesting to go back – either in your imagination or in real time – and experience the emotions your childhood home triggers. You can discover much about your childhood and the way you were brought up, how content you were or whether there was a lot of stress. Your home may seem smaller than it is in your imagination but the memories and sensations will tell you a great deal about where you come from and the beginning of your journey.

Going back to the family home

What are your memories of your childhood home? Was it a sanctuary or a place to escape from? Was it warm and user-friendly or stiff and formal? Could you find your own space or were you all on top of each other?

- Imagine your childhood home. You can draw it if you like or do this as an imaginary exercise.

- Go in through the front door and walk yourself through the rooms. Feel your way back in time and sense the emotions in each room. Where did your family congregate? What was the atmosphere like? Imagine family gatherings at the dinner table. Were they happy or fraught? Take yourself out into the garden or yard and feel how you felt back then.

This is the place where your parents created a foundation for their new family. It will have reflected their thoughts and feelings about themselves and their children. How has this affected you in your own life?

SEVEN
In the Shadow of the Ancestors

There are some trees, Watson, which grow to a certain height and then suddenly develop some unsightly eccentricity. You will see it often in humans. I have a theory that the individual represents in his development the whole procession of his ancestors, and that such a sudden turn to good or evil stands for some strong influence which came into the line of his pedigree. The person becomes, as it were, the epitome of the history of his own family.

SIR ARTHUR CONAN DOYLE

We carry the imprint of our ancestral genes in our DNA, which will determine the way we look, as well as some of our character traits and physiological disposition. We will inherit some that is good and some that is less positive.

Our physical heritage may be obvious in the way we look, walk and talk – our gestures and the shape of our hands and feet are often carbon copies of our parents and grandparents – but we also inherit more subtle influences, such as hidden values, beliefs and attitudes that have been passed down through the generations. Some of these traits are the result of our environment when we are growing up, but genetic factors also predispose familial groups to certain types of behaviour.

When we recognize that some aspects of our behaviour have been passed down to us in our ancestral heritage, we can begin to look at ways of changing the negative expressions of ancestral memory and building on the more positive aspects of our inheritance.

On the dark side

> We all grow up with the weight of history on us. Our ancestors dwell in the attics of our brains as they do in the spiralling chains of knowledge hidden in every cell of our bodies.
>
> **SHIRLEY ABBOTT**

Ancestral disorders that are passed to us through our genes and memories can overshadow the family. Some of my clients recall that they began to develop symptoms of distress in their early childhood. They remember sleep disorders and night terrors, anxiety, that led to migraines, panic attacks or asthma. They were often the sensitive member of the family who picked up on unspoken anxieties and family troubles.

Philip, for example, became depressed after discovering his alcoholic father unconscious outside the family home. He witnessed the physical abuse of his mother by his father, which triggered a cycle of childhood depression and an anxiety about going to bed. When it got dark, Philip was ready with a thousand and one ways of counteracting his fear of sleep. 'As soon as I fell asleep there was this overwhelming darkness, this heavy uncomfortable abusive feeling that came up from deep inside my body. I thought it was trying to attack me. I frequently had nightmares and whenever I was sick with a high temperature, I used to hallucinate and occasionally sleepwalk. My dreams were always in colour, imaginative and extremely vivid. As an only child, I had no one to share my fears for my mother or to understand the heaviness that my father brought home with him when he had been out drinking.'

Philip is now married with two children and his eldest son suffers the same syndrome. 'Since he could talk he would tell us his dreams and some nights would be anxious about going to sleep. His nightmares were always to do with being suffocated by a heavy oppressive feeling, and he's still afraid to sleep without his light on. I have since discovered that in my father's family there is a long line of alcoholics and abusive personalities and whether this is a manifestation of their shadows that affects my son I don't know, but as a result it comes out in bad dreams and a deep-set anxiety of feeling unsafe.'

Children's senses are more attuned to emotional and physical atmospheres than those of adults, so if we were unfortunate enough to experience violent behaviour, physical or sexual abuse, rejection or abandonment, it is worth knowing that these circumstances can cause children to become depressed and anxious, to have nightmares, behave rebelliously, self-harm or develop eating disorders – in most cases as a way of suppressing their unhappiness and helping them to cope with family toxicity.

During childhood we can 'learn' or adopt psychological disorders from our parents. These may not be dealt with until we have become adults and are going through a crisis for which we seek some kind of therapy. Perhaps only then do we begin to recognise that our compulsive behaviours, fears and anxieties actually stem from our childhood. Even if we understand these patterns, it takes great courage and determination to change direction and find ways to control or overcome these influences.

When children develop strange habits and phobias this can be a clue to a family memory. As a child, Sharon had a fire phobia: she remembers sitting up in her bed every night, checking to see if there was smoke blowing under the door. 'My big fear was that the house was on fire and I had to find a way to escape.' History repeated itself when her three-year-old son expressed the same fears. 'At a family gathering, my aunt admitted to having an identical phobia. My aunt went on to say the story she had heard when she was young, how a sister of her father so my great-aunt and her whole family, apart from one child, had died in a house fire. She believed this tragedy to be the root of the family phobia.'

When a family is dysfunctional, the atmosphere can become quite toxic and unhappy for children living in those households. This term can also be applied to families where life is not quite right, causing harmful attitudes to sex, drugs and abuse. If we can see how these dysfunctions have affected us then we can see how they might be affecting our own children and we can begin to deal with them. Step by tiny step, we can address them without prejudice or fear and so manage what is inherited from our family.

Psychological disorders and mental illness

A whole range of different factors can make us vulnerable to mental illness, from our genetic blueprint and brain chemistry to lifestyle and past experiences. Many of us can look back into our family and find patterns of dysfunctional behaviour. By digging a bit into your own family history of mental health issues or disorders, you can begin to be aware of these patterns and identify the suffering individuals in your family's ancestry.

I see clients who suffer from regular bouts of depression and come to recognise a link to psychological patterns in their family. Sophie found out she had inherited depression when she had a nervous breakdown in her late twenties. She told me, 'When my boyfriend left me I collapsed into a familiar cycle of anxiety and depression, which in this case just got worse and worse. Then I discovered that my father too had suffered from depression when he lost his job; my mother hid it from us so nobody knew he was sick.'

After Sally was diagnosed with depression she was not surprised to find that many of her family members had suffered mental health difficulties. Her maternal grandmother was hospitalised for more than a year with debilitating depression during the 1950s, when mental illness was shameful and not talked about. She then discovered that her great-grandmother was a pyromaniac, and was sectioned after burning down the family home. Her grandmother's father on her father's side had killed himself decades earlier. Her father's siblings had both committed suicide in mid-life, not long after her own diagnosis of depression, and her father was then diagnosed with bipolar disorder.

Sally said, 'Uncovering my family history has not stopped me seeking ways to manage what I have. I find that this biological self-knowledge can be empowering and, since my genetic journey has begun, I can say that it has given me an understanding of myself and I now know that I react more strongly to negativity than other people and I treat myself with a little more care and understanding. I realise that I am not completely responsible for what I am carrying – it's in my genes.'

Depression is one of the most common inherited disorders and scientists have for the first time identified a gene that controls the production of serotonin in the brain (low serotonin levels are associated with depression). This is a real breakthrough, but having this gene does not make it inevitable that the individual will suffer from depression, because serotonin levels are not its only cause.

For some people depression comes and goes, depending on stress levels and difficult circumstances such as bereavement, unemployment or a break-up, but it can also affect many people who have no apparent precipitating event. Studies suggest that, in the population as a whole, 40 per cent of cases of depression can be explained by genetics and the rest is related to environmental factors. Bipolar disorder and schizophrenia are also known to have a genetic component, but as yet it has not been possible to pin down the precise genes that are involved.

Sometimes clients will search back into family history and find a parent, grandparent or great-grandparent who committed suicide. Most become fearful that they too will not be able to cope with their depression and that it could lead them to their own suicide. This is a very real fear but with the right treatment their depression can be managed.

Melissa, an heiress from Scotland, has been suffering from depression since she was seventeen years old. She told me that her grandmother, a bright society star during the 1920s, had suffered from depression and chose unhappy relationships. Her mother also suffered, and the family secret is that she did commit suicide. Melissa's upbringing was one of privilege and wealth but she found that money and status was a double-edged sword.

'I realised that title, money and beauty do not equate to happiness, and actually the higher you climb the more difficult it is to find happiness. As a family we were not really emotional and there was no affection between my parents. So I come from a line of lonely and unhappy women who, on the surface, appeared to have everything. Now I am married and have two children I want to stop repeating these patterns and have sought help from a therapist to heal the loneliness I suffered during my childhood, which pushed me into my addictive behaviour and depression.'

Science has also established that there is an addictive gene. There is a 70 per cent chance of becoming an addict if you have the gene along with environmental factors such as addictive behaviour in family members. Addiction is a learnt response and a way of coping with life.

Whatever the addiction – from drugs to compulsive internet use – addicts believe they are better equipped to cope with life because of it. Addictions may be part of our inherited genetic make-up, but just like any ancestral shadows they are deeply rooted within our subconscious. When addiction or abuse runs in the family, is it possible to break the vicious cycle?

Some of us are born with the addictive gene, but its influence may remain latent until traumatic circumstances such as separation or death trigger the actual addiction. A person who has been brought up with parents who are addicted will have a nine out of ten chance of becoming an addict themselves. It is a combination of brain chemistry and lifestyle, so you are either drawn towards the addiction or away from it. The tendency is to be an alcoholic or not to drink at all. There are no half measures.

Christine's family circumstances triggered her addictive behaviour. When she was recovering through therapy, she looked into her family history and saw that she had inherited both her mother's and father's gene pool of destructive and negative behaviour. Her mother was a depressive and her father and grandfather were both alcoholic and abusive.

She was sent to a children's home when she was thirteen because her mother was severely depressed and couldn't cope, and she in turn was forced to send her own children into care after leaving her husband – himself a violent alcoholic. 'Not feeling wanted runs in our family,' she says and that feeling was handed down to her daughter, Sam, who was sexually abused by her father and by the carers in the home before being sent to a loving and stable foster home.

But at fourteen Christine discovered alcohol and it gave her confidence. 'I started drinking on a regular basis,' she recalls. 'Then I became anorexic and the problems continued until I met a former alcoholic who wanted to help me and I was admitted to a clinic where he was working.' She found the strength and courage to work through her own and her family's pain and eventually, after years of sobriety, she trained as a counsellor and now works at a clinic herself.

Katherine Hooker, a fashion designer, discovered that her family's addictive personalities stretched all the way back to General Joe Hooker, the famous American Civil War general who had a reputation as a hard-drinking ladies' man. A great-uncle was one of the founder members of AA and her great-grandmother lost everything in the crash of the 1930s which traumatised the entire family.

Katherine found a way to overcome a low-level eating disorder that had bothered her since childhood. 'For a long time I found it was so easy to get wrapped up in family issues so I didn't actually get any better,' she says. 'Now I know that my life is whatever I want it to be. I have the power to create my life and my family is not going to be the excuse for me not to do that. I have the freedom to choose. But before I could reach such a positive understanding I needed to look into my family background.'

Once we see our negative ancestral influences, we can begin to understand how these negative self-expressions manifest in our own lives. Looking back through the family tree, we can trace our susceptibility to physical and psychological problems and disorders. Finding ancestors who were affected by addictions and dysfunctional behaviour will tell you whether you might have inherited the potential for these disorders. If you have, then recognising the risk factors which may influence your behaviour will help you to overcome or manage your inheritance.

Violence and abuse

> The concepts of human behaviour being derived by either genetic factors or by environmental factors is no longer tenable . . . in most cases, both factors are vigorously at work, the only question is to what degree.
>
> **DR GLAYDE WHITNEY**

Just as dysfunctional personalities can resurface through the generations, so can criminal and pathological behaviours. These disorders are known to run in families, often skipping a generation or two before manifesting again, perhaps triggered by family trauma or abuse.

Past traumatic events in our family history or a cycle of repeated abusive or dysfunctional behaviour passed down generation after generation can create darkness in certain families. People who are exposed to this behaviour can respond in different ways: they may defend themselves from the abuse by denying it, or they may react with unemotional aggression, showing no remorse or awareness of the way their behaviour hurts or wounds other people. If we have ancestors who have died suddenly or traumatically through murder, suicide or tragic accidents, they can send reverberations through the family tree. And they may be the key to our deepest, darkest aspects, influencing our behaviour as their living descendants so that we continue passing on this negative inheritance.

Factors such as poor parenting or abuse can increase the likelihood of the inheritance becoming active in the next generation, while a happy childhood will reduce it. Someone who is surrounded by criminals as a child is more likely to become a criminal themselves, and abusers do tend to have been abused. This can lead to a cycle of repeated abuse over generations. Many women who have seen their fathers abuse their mothers seek abusive relationships themselves, which continues the family pattern. The abusive behaviour seems to get stuck in an endless loop.

Mark, a police officer, has worked with families whose relative or child has been a victim, an abuser or a murderer. From his experience, 'The most common occurrences of violent behaviour happen at home, such as with domestic violence and incest. Most offenders want to control their victims and many do so through this type of violent and abusive behaviour, which then continues to affect the family by association and learnt behaviour by their children and children's children. Often an offender when caught will completely deny that they committed the crime, some actually convince themselves that they never did it.'

This may explain how a violent act such as murder or abuse can become a hidden secret within a family, a trauma whose legacy is only realised several generations later when it manifests as an ancestral shadow.

Jenny is one of five children and her eldest brother is now in prison after murdering his wife and child in fitful rage when his marriage was in trouble. Jenny remembers: 'We knew he was different when he was little, he just never knew when to stop; he was intense, and really insensitive to anyone else except for himself. When my parents found out that he had killed his wife and their grandson, they were both shocked and devastated. I think that as a family we will never recover but I did wonder if his tendency was inherited. I found out that my grandfather on my father's side was a difficult man. He had a tendency to plunge into violent rages that frightened my father. He said that when my brother was born he had a look of his father.'

The most famous UK perpetrators of organised crime were the Kray twins, Ronnie and Reggie. They were famously involved in criminal activity in the East End of London during the 1950s and 1960s, along with their elder brother Charlie. Their father was a scrap gold dealer, frequently away from home, and he then became a deserter during the Second World War, so he was either away or hiding from the law.

The boys were always into trouble at school, and by the time they were sixteen they were notorious. They were found guilty of murder in 1968 and their history of violence and criminal activities followed them into prison. Ronnie was certified insane and died of a heart attack in 1995 in Broadmoor, a high-security hospital, and Reggie died of cancer a few years later. Charlie became a drug smuggler but was eventually caught and sent to prison, where he too died. The family legacy continued: in 2008 Charlie's grandson Joseph and his son-in-law Norman Jones were convicted of torturing, killing and beheading a man they accused of stealing drugs that belonged to them. They have been given life sentences and history is repeating itself.

It is believed that the souls of perpetrators attach to their family line, especially if the murderer who committed the atrocity never repented the crime: it then becomes a family legacy.

Edward Tick, a therapist who specialises in PTSD, told us a story of a Vietnam veteran who had killed innocent civilians during the war. His family built the war orphans a house in Vietnam, as a Christmas gift. The children and the grandchildren were atoning for their relative's behaviour. Many cultures around the world have rituals for making peace with the ancestors who are still hanging around, whether victims or perpetrators, offering them a legacy of peace and forgiveness.

Jane has worked in psychiatry for over twenty years, nursing in a variety of establishments including maximum-security prisons. The majority of her patients have a criminal background and she has nursed a very high number of murderers and rapists. Jane believes that she may be healing a family tragedy through her work. She told me, 'There is no history of any family member working in my profession. I do not know what drew me to do this work, only to say that human behaviour fascinates me. I am the youngest of four children and the only girl. When I was born, I was called Jane. It was not until I was in my very late teens, when I had already started nursing and my sexuality was established as being gay, that I learnt that my father had an aunt, also called Jane, who was raped and murdered in a local beauty spot. My mother had no knowledge of this story when I was born and named me. I am drawn to my work unconsciously, possibly to heal what happened to my great-aunt Jane.'

We can inherit good as well as evil and most of us are somewhere in between. We could think of human good and evil as a kind of continuum on which our place can change, for better or worse, in as little as one generation.

Physical health

> A further sign of health is that we don't become undone by fear and trembling, but we take it as a message that it's time to stop struggling and look directly at what's threatening us.
>
> PEMA CHODRON

Our bodies are a perfect reflection of our inner consciousness. How we feel emotionally, how we think, what we have inherited, each organ, gland or part of our body reflects different aspects of the psyche and when a part of the body is diseased or damaged that damage can stem from a long line of ancestral influences. Illness can bring us into a recognition and understanding of aspects of ourselves that we are holding on to too firmly. Sometimes it helps us see the changes that we need to make, so healing can be seen as a gradual awakening to reality.

Clare is a creative director from Swindon. 'I had always been a high achiever until I became really sick with glandular fever and for five years I was bedridden with ME [myalgic encephalomyelitis or chronic fatigue syndrome]. At the same time my maternal grand-mother also became really sick, so she came to stay with us. I found out lots of things about her life that I never knew. She too was a high achiever but her father never paid for her to go to university, so she married someone who she didn't really love to get away from the family home. She told me that she always felt angry and disap-pointed by her life and felt that she had been affected by her mother, who was rather distant and controlling.

'During the time I spent with my grandmother I talked to her about my dreams and ambitions. I believe that through our chats she encouraged me to get better, to follow my heart and do what I wanted to do. She told me that the time for opportunities is when you are young and full of hope, and that this hopefulness disappears when you get older and you can lose any belief in your dreams. At that moment I decided I had to break the patterns of my mother and grandmother, that I would find a way to get better and live my life. I am now a creative director with a well-known international chil-dren's charity.'

When we feel ill, the sickness may be spiritual rather than

physical. Accepting this is an important step in the healing process. At this point the symptoms are released and the deeper levels are addressed as well.

Healing the mind, healing the body

> They were Eastern European Jews and immigrants into Britain and had been decimated generation after generation. They were ethnically cleansed by the Mongols and the Russians and we hold that memory in us. And that memory manifested in my body in the form of these tumours.
>
> **DAVID SYE**

Ancestral issues arise in many different ways, but they can manifest in ill-health as our bodies respond physically to the pressure of old patterns and family dysfunctions.

David's life came to a crisis point in his late twenties when he was diagnosed with ulcerative colitis and a spastic colon. 'The doctors found twelve tumours in my colon. I was in near constant pain.' They told him he would have to take painkillers for the rest of his life. David instead opted for Tibetan yoga. After just ten sessions he went back to the doctors, who couldn't find any tumours. 'I felt better than I'd ever felt before.'

David began practising yoga and exploring other avenues of healing and faith. He also broke away from his family and the shadow cast by his famous father, Frankie Vaughan. 'My body had reacted to my holding on to the past. As the eldest son of a Russian Jewish family, I was trying to be what my clan expected me to be and it was making me sick. They didn't mean it, of course, but I was born, like everybody else, to highly infectious adults who were full of emotional poison and that poison was passed to me.' For ten years David travelled the world doing a variety of different jobs and carving a new path for himself. Two years before his father died, he got back in touch. 'We resolved everything that we needed to.'

Today he is a world-renowned yoga teacher with a gritty, unorthodox approach who has used his classes – practised against a background of hip hop and world music – to bring together Palestinians and Jews in Israel. He also works with prisoners, drug addicts and deprived youth. 'I have no blame for what happened in my body as a result of their legacy,' he says. 'It woke me up!' By taking a proactive approach to healing, David discovered his life path and in the process healed his relationship with his family and his ancestors.

From a holistic perspective disease always arises from a wider imbalance; literally dis-ease in the family, in the community, with the spirit world and the ancestors. Holistic healing recognises the entire system of mind, body and spirit. Medical intervention may be necessary and pharmaceuticals advisable but it is empowering to consider other issues behind illness and to work in non-linear ways to alleviate it.

Mercedes from Gran Canaria is one of thirteen siblings, many of whom have died from cancer. As the daughter of a doctor and medically trained herself, she is not one for assuming that there could be a mystical cure for her siblings' illness, but when three brothers were diagnosed in as many years she felt there might be something other than physical illness going on in her family. At one of our retreats, Mercedes told us that a curse may have been put on the family. She had discovered that her family origins went back to the time when the islands were conquered. An ancestor had been forced to marry one of the conquerors and it was believed that a curse was placed on the family, affecting each subsequent generation. She asked how she could perform a ritual to heal the family.

Mercedes and her niece, Peggy, the daughter of one of the brothers who had died, decided to perform a ritual together. First they went to the original home where the family had lived, the church linked with the Virgin Mary, goddess of the island of Gran Canaria where they had worshipped, and the graveyard where her family were buried. They asked the Virgin Mary for her blessing and then, when they had connected all three locations, Peggy and Mercedes created an altar for their living and deceased family members and lit candles and placed flowers on it for three days. They needed to link the history of the past with the journey of future descendants so that history did not repeat itself. 'Peggy and I completed our ritual by taking flowers to the graves of our deceased family and ancestors and felt this enormous relief as we asked them to forgive what had happened and we also promised that they would never be forgotten.'

Now her brother is in remission and a sense of possibility rather than loss surrounds the family. They have been celebrating the arrival of new additions to the family, and new births bring new hope.

The origin of family illness may be difficult to locate but searching out the roots of dysfunction gives us, at the very least, a sense of empowerment. We are no longer powerless in the face of what can feel like insurmountable odds. Sometimes we are able to heal a condition completely but even if this is not the case there is comfort in connecting with the bigger story of who we are and where we have come from. A spiritual relationship with our ancestors can help us find the inner strength to cope with illness and teach us to feel safe even when we are unwell.

But we also need to face the physical inheritance from our family. When I work with people, I can see generations of family pain in their bodies. They carry it in their faces, in their bones, in their flesh and the structure of their being. Different aspects of the body carry different aspects of ancestral heritage. For example, the womb is always connected to the female lineage and the heart contains the eons of ancestral grief and hardship that our ancestors endured. Meanwhile, digestive disorders reflect the results of the fluctuations of feast and famine that our ancestors experienced. As our bodies carry latent memories of family history it is useful to find ways of releasing those memories. Acupuncture, cranial sacral therapy, shiatsu, even deep tissue massage help our bodies release the memories they are holding, whether they are from our own lives or passed down from our family.

In my physical therapy, I first sense the client's connection with their mother, reflected by energy on the left-hand side of the body. Starting from their feet, moving upwards through their pelvis, solar plexus and heart, reaching up into their shoulders and head, I can sense how much physical comfort they received as children. When this side of the body is relaxed and balanced, it reflects a strong and loving relationship with their mother. The same happens when looking at the right-hand side of the body as a reflection of their relationship with their father. We can then look deeper into the inherited issues and how that has affected them. Untying these knots begins with letting go and forgiveness, and that is when the body starts to heal.

Shamanic healer James Hyman (founder of Deep Emotional Release bodywork) explains, 'We're emotionally based but we've learnt to suppress or ignore our emotions. We need to do as we did as small children and release all this pressure by weeping. When we go to that place as adults we release our personal story and at that moment the ancestral struggle and pain is released.' In cathartic bodywork sessions, James helps people release the traumatic memories held in their bodies, which in turn frees them from the past and allows them to be more fully in the present moment. He says that in liberating themselves, his clients are also liberating their ancestors. 'When an individual heals it within themselves, when they disassociate themselves from the ancestral story, from the "my mother did this, and my father did that", the ancestral chain is healed. We are, in effect, healing the DNA.'

Therapist Anita Bains has witnessed the complete alleviation of symptoms in victims of abuse, addiction and extreme PTSD using Emotional Freedom Technique (EFT). 'Our body remembers what our mind forgets,' she says. Anita believes this mind-body system can heal core ancestral issues that begin when we pick up our mother's emotional and mental state whilst in the womb. First she finds out what someone's programming might be and then helps her clients to release the energy pattern associated with it. 'It shows up in many different ways: depression, unhappiness, all the addictions, relationship problems, anxiety and also in physical illnesses,' she says. 'Every physical illness is the result of a disturbance in our energy system – and many of them begin with our thoughts and our beliefs about ourselves. Auto-immune diseases, such as chronic fatigue syndrome, fibromyalgia or lupus are directly connected to lack of self-esteem and a general feeling of unworthiness that can be passed down without our being consciously aware of it.'

Healing is always multi-faceted and we can use as many different forms as we are drawn to: David cured himself using Tibetan yoga and then set off on a long journey of self-discovery before being able to return to his family fully healed and fully present in who he was; Mercedes used ritual to help heal the physical illnesses that were affecting her family.

Working with a combination of physical release work, psychic transmissions and spiritual healing means that the ancestral hardships, trauma and pain lodged in our body and emotions can be released. Then we are able to see more clearly who we are, without ancestral shadows, and lead our own lives rather than following the programming of our family.

The creative soul

> Follow your bliss and you put yourself on a kind of track that has been there all the while, waiting for you, and the life that you ought to be living is the one you are living.
>
> **JOSEPH CAMPBELL**

When I was pregnant with Ossian, my first son, I had a powerful dream: I was in the Grand Canyon and an American Indian elder came to me to tell me he was giving me his son. I didn't really think much about it at the time, but as Ossian grew up he began to display a preternatural connection with the natural world, which is at the heart of American Indian heritage. He is dyslexic, which has made conventional study a real challenge, but he has a bond with animals and an understanding beyond his years about ecology, nature and the world around him. This is his gift. And, while I know of no ancestral link with American Indians, I believe this dream was a guide from the great pool of our ancestral human heritage, encouraging me to recognise my son for what he is: a brilliant natural ecologist.

Traditional cultures believe we are all born with a special gift, and when a mother has a dream like this the family and community take it as a message from their ancestors, telling them who is being born into the family and what they have come to do. At the birth of the child he or she would be given a special name to reflect and protect who they are throughout their life.

We all have a unique purpose and we are born with the tools to realise that purpose. In ancient Greece, the abilities of each and every human being were attributed to their soul, which was known as their genius. So in that sense, we all have genius. The important thing is that we realise the innate potential that taps on our consciousness until we decide to listen to it.

The creative gift

If you want to work on your art, work on your life.

ANTON CHEKHOV

Each of us is born with our own creative spirit that is nurtured or hindered by our family and our environment. Our creative personality can show itself as early as two or three years old in a special aptitude for language or music, precocious mathematical ability or manual dexterity; it usually surges again between the ages of seven and fourteen, a time when we start to express our personality in the world.

In the best scenario, our family supports and inspires our gifts. Dr Wendy Denning has a successful private medical practice in London. Her inspiration was her grandmother, her mother's mother, a lifelong diabetic who became one of the first experimental patients to use insulin to control her condition. She was diagnosed when she was eleven years old and eventually died at seventy-five, only the third person in the world to use insulin.

As a result, Wendy grew up aware that medicine had saved her grandmother's life. It propelled her towards becoming a doctor using integrative and complementary care. 'My grandmother was interested in health and spiritual subjects. She was brought up in a home where service to the community was important, being the daughter of the Chief Justice of the United States. When China opened up to the West she was one of the first women to go there, and this led me to explore Chinese medicine. To this day I still feel her driving me to support the community and seek solace in spiritual subjects.'

Unfortunately, we can be easily influenced by the expectations of others, by peer pressure or by family rules and values. It is then you may find that you have lost your connection to your calling and your innate talents. As our parents push us into professions that appear to be more lucrative or stable, we may lose sight of our creative gifts, but they never go away. They are always there waiting for us to wake up and follow our heart.

By looking into our historical past, we may find characters that inspire us. Gifts and talents often skip generations, so while our living relatives may not display any interest in our interests, or seem positively opposed to our chosen direction, we might find a soul accomplice further back in our family line.

Ben had no idea what he wanted to do on leaving school, but after several weeks of work experience in a jeans shop he knew he had to begin to study and push himself academically. 'I began to focus on what I really wanted to do and realised that medicine really inspired me. It was strange, as I did not grow up in a medical or science-orientated household.' A few months later, he found out that his great-grandmother had been a nurse and his grandfather on his father's side had been a GP. Ben is now a doctor.

By looking back into our ancestry we may find a character or personality who seems to resonate with our deepest dreams and desires. Or they might simply be someone who rebelled against the conventions of their time and forged their own path. But whether or not you can actually pinpoint one particular ancestor, you can be sure that they will be there encouraging you to manifest your passion. You were born to fulfill your creative destiny and just as there may be ancestral shadows working against you, you can be sure that there will be a benevolent energy supporting you. You are, after all, a part of the legacy of the generations who walked before you.

The creative dynasty

Some of us are lucky enough to be born into a family where our innate talents and unique personalities are naturally encouraged and fostered by those around us. Then there is a perfect match between our genetic heritage and our environment.

Eliza Carthy, a multi-award-winning singer and musician, first joined her parents on stage when she was six and formed her first group, the Waterdaughters, with her mother, her aunt and her cousin when she was just thirteen. Carthy's family is steeped in music tradition and folklore. Her parents, Norma Waterson and Martin Carthy, are folk royalty and, recognising early talent in their daughter, encouraged her in every way they could. 'There's a strong sense of continuity in our family,' she says. 'We believe that everything is a big circle and although my father has Irish heritage, I do feel a strong English soul working through me in my music.'

She can trace her musical roots back seven generations to Thomas Carthy, known as King of the Pipers, who was born in Ireland in 1798. Her mother's side had gypsy blood and all of them had musical talent. 'My grandfather played the trumpet in the pit at the cinema; another used to dance on the bread board so there was a family joke about my grandmother's bread always tasting different. And when another granddad was in the Special Forces he used to play the banjo to all of them.'

She is proud of her heritage and spent a large part of her early days sticking up for folk music when it seemed no one was interested. And so Eliza, standing on the shoulders of her folk giants, forged a new path that has resulted in a revival of nu-folk artists like Mumford & Sons, Noah and the Whale and Laura Marling. And now she is passing on her musical gifts to her daughters Florence and Isabella.

There are vaudeville families and circus families; families who are brilliant entrepreneurs and families who are all chefs; families who go into the law and those who are all criminals. Sometimes professions change over generations, so a family in which one vicar followed one another may change direction but their descendants may still reflect their heritage in their choice of career, perhaps showing an interest in religion or some kind of pastoral community care.

There is always one ancestor who starts such a lineage. Emilio and Marco Nella are the biggest knife sharpeners in Britain, supplying freshly sharpened knives to over 12,000 customers from high-scale London restaurants like Carluccio's to supermarkets like Tesco. Their great-grandfather began it all in the village of Carisolo in northern Italy. When he was twenty-one, he set off across Europe with his knife grinder in a wooden barrow and, sharpening knives as he went, arrived a year later in Deptford. His sharpening block still sits in Nella's showroom. Following his example, other members of the family emigrated and went to start knife-sharpening businesses in Brooklyn, Philadelphia and Montreal. There is now a statue of a knife sharpener on the outskirts of picturesque Pinzolo, a neighbouring Italian town, as a tribute to all the knife sharpeners from the area.

Sisters Sue and Abigail Dooley have both been really successful writing and producing television and theatre. They both have brilliant comic timing, an instinct for good ideas and an admirable ability to get things off the ground, thanks to their father, who worked with the comic geniuses of the sixties and seventies in British television.

As Abigail says, 'Creativity is in my DNA. The driving force comes from my father and my grandfather, both creative geniuses, highly inventive and original.'

Her ancestors were part of the famous circus that featured Joseph Merrick, the Elephant Man. Her great-grandmother was Fatima, the Bodyless Woman and her son and his cousin were billed as boxing midgets, although they were no more than children. Her grandfather became a magician and her father joined the circus at six years old with his three sisters. He eventually struck out on his own as a stand-up comedian and magician, writing comedy for television and finally producing light entertainment programmes.

Their parents encouraged the girls to get out in the world and make a living from their creativity. 'My dad taught us the meaning of hard work when he would stay up all night working on material for a recording the next day. The best advice he gave us was "have the courage of your convictions".' Sue and Abigail have been given the perfect combination of a genetic inheritance that was nurtured and supported, and consequently both girls have flourished in the notoriously competitive world of theatre and television.

Parental support is a key element in achievement, as Sigmund Freud acknowledged when he said, 'If a man has been his mother's undisputed darling, he retains throughout life the triumphant feeling, the confidence in success, which not seldom brings actual success along with it.' Comedian Eric Morecambe had no theatrical background. His father was a council road digger, and his mother a waitress. But his mother Sadie recognised his gift for performing and making people laugh, and entered the young Eric for talent competitions. 'Eric was her only child, and she wanted the best for him,' says his son Gary, a writer. Eric was the first one in his family to break away from his background and forge a new template that has opened wider horizons for his descendants. Gary's children are all drawn to music and the arts, especially Arthur, the comedian of the family, of whom Gary says, 'He's inherited so much of his famous grandfather's character and talent.'

If you are a parent, support your children in their creative endeavours, no matter how different their interests may be from your own. Even if you have no prior family history to draw on, you may be the inspirational ancestor who lights a creative spark which your children will hand on to future generations in their turn.

Behiye's story

Singer-songwriter Behiye, a first-generation Australian with Turkish parents, has taken an epic journey to discover her true voice and embrace her heritage. At first she found herself having to push against her parents' traditional ways. They might have given her a western education, but they really wanted her to marry a good Turkish boy, settle down and start a family as soon as possible.

Behiye's artistic spirit rebelled. She refused to conform, wrote poetry and travelled widely. But in her mid-thirties she became discouraged. Her dreams were not manifesting, she was not finding the ways she wanted to make music, she was disappointed in her relationships and, finally, a business deal fell apart, leaving her with no option but to return to her mother's house.

Despite her rebellion against her mother's traditional Turkish ways, Behiye had always felt drawn to the culture, the food and the music of Turkey. So she decided to go and live there for a while. Everything opened up for her: she met her music teacher and gathered members of her new band within weeks of arriving. And her world has changed in other ways as well.

'A very important part of me has been missing for years and I have found it again. You can't run way from who you are forever. I have made my peace with my cultural heritage and this has strengthened me as a person. I have been welcomed both as an artist and by my family. I think you have to know where you come from to know where you are going. And Turkey has helped me know what tribe of song-writers I belong to. In western music, they are called folk artists, in Turkish music these songwriters are called *ozan* or *asik*. They have a message, they tell a story, and that is what I do.'

Behiye needed to make that link to her deep roots in order to find herself and her music: her rebellion gave her the strength to pursue her passion and break away from the more controlling aspects of her culture, but it was the country of her ancestors that gifted her with her true voice.

A soul legacy

There are those in each family who embody the soul of the family, either consciously or unconsciously. It is as if all the light and the dark of their ancestors' experiences have culminated in them. Carla Esteves, a women's health practitioner, doula and healer, is one such person.

When Carla looked back into her family history she began to understand her spiritual connection with all the women in her family. 'As I uncovered the story, I understood what gifts I had for the world.' She changed from studying ecology to becoming a healer.

'Naturally I have ended up working for women, attending their births, helping them during breastfeeding or seeing them for womb healing. I supported women in traumatic situations like miscarriages, abortions, birth and menstrual pain, listened to women's feelings and the emotions of their disconnection from their femininity. I found myself offering them a way to work through that by connecting with the earth and with their bodies.

'As I delved into the lives of my great-grandmother, grandmother and mother, I learnt about the stories of their bodies and hearts, of their lives, their men and their children that shaped them as women.' Carla understands the effect her heritage has had on her. 'My heritage is influenced by the country I was born into and the places that I have lived, and despite the physical distance between my grandmother, aunt and cousin in Portugal, my mother living in Spain, and myself in England, I can still feel the grounding, nourishing, unspoken strength of our bond, in the same way as I did as a child when we all shared the same house in the city of Porto.

'Like a constellation in the sky, the image that usually comes to me when I sense the configuration of this relationship is that of a six-pointed star, the sixth point being my great-grandmother, who passed away when I was ten years old, and who is my guardian ancestor and my guide.'

Katherine Hooker created her own successful fashion label after she kicked the self-destructive addictive behaviour that had trammelled her family since Fighting Joe Hooker. And as her business took off in America, she discovered a whole host of ancestors who now inspire and reinforce her singular vision and commitment. Thomas Hooker is a particular inspiration. The breakaway Puritan leader founded Connecticut with the words 'the foundation of authority is laid in the free consent of the people', that became the fundamental principles of the American Constitution. He is still celebrated as the Father of Connecticut and is honoured on Hooker Day in Hartford.

'Where once it seemed quite daunting to launch my business in America, now it doesn't feel that way. Discovering these ancestors has been really reassuring. They are a kind of fuel that keeps me going and I do feel that I am honouring their memory in the work that I do.' She is the embodiment of their legacy as her own small empire grows exponentially and with it her ambition to empower everyone, from weavers in the Hebrides to her staff in London and New York and the women who wear her clothes.

When we connect with our family tree, we become the link between past and future. By understanding the journey of our ancestors – both those who succeeded and those whose lives were troubled – we are aware of our collective evolution. Honouring the journey of your ancestors by gathering all the gifts they have given you creates a legacy that will sustain those that follow.

The blocked creative

Many people lose their creative spark and cannot retrieve it. As children we have few of the filters that protect us from criticism and ridicule and, unable to overcome the programming of our parents or the education system, many of us quietly shelve our dreams. Over time this can cause problems such as depression, since our natural creative impulses need to be expressed.

Our creative blocks can come from many sources, including being overshadowed by an extremely successful parent or sibling or having our gifts suppressed in childhood. More often than not, this suppression is mirrored in previous generations and is passed down by unhappy, unfulfilled parents and grandparents whose creativity has been thwarted by circumstances. They may even be jealous of the younger generation's creative opportunities.

When Michelle was young she desperately wanted to go to art college but her father refused to let her go and sent her to university to study languages instead. She has been searching to find her creative impulse ever since and has suffered from ME and depression for years. When she first attended art college about ten years ago, she felt unable to draw or paint. 'It was as though a part of me was paralysed,' she explains.

However, Michelle refused to give up. She noticed that when she was doing something she enjoyed, her depression and fatigue disappeared, so she decided to try once more. 'I had the courage to enrol at a local college and have finally found my passion in painting watercolours.' She later found out from her sister that her great-aunt was a well-known life-drawing teacher in a small town in France.

It is never too late to retrieve our own creative path. But it is necessary to look at areas of your life that you are unhappy about. Are you reflecting the same issues as your family – both learnt and inherited? Are you expressing your true spirit? Did your parent, your grandparent or great-grandparents follow their dreams? Or were there circumstances that created restrictions, failures and lack of opportunities?

The process of exploring our family's roots and identifying these influences can have a tremendous value in terms of gaining a deeper understanding of ourselves. It is impossible to free ourselves from some of our childhood conditioning, but we can learn to accept what we cannot change and focus on changing what we can.

If there are positive qualities that you share with your family, look to see what action you can take to encourage and shape these gifts in your life. Or if there are any negative attributes, look at what small steps you can take to change them. For example, if there are workaholics on both sides of your family and you are a workaholic, you might make a commitment to do nothing for an hour once a week. If your family procrastinates, then make sure that you do one thing on your to-do list towards furthering your creative pursuits every day.

Many ancestors were unable to fulfill their own dreams, and those dreams live on as shadows of unfulfilled desires in their descendants. What they leave behind when they die can become a burden for a descendant whose personality responds to the unfulfilled dream. Your ancestors' stories will help you recognise which of them led unhappy lives and which fulfilled their destiny and left a positive legacy.

To release the non-creative ancestors, write them a letter about your life changes. Thank them for all the support and care that they have shown you, telling them it is now time for you to go forward without them. Ask them to release their personal desires and restrictions. You can write to the ancestors you believe could be blocking your creativity or those you know had difficulty in fulfilling their creative dreams, and if you don't know who they are, you can just call them your non-creative ancestors.

Once you have done this, you are ready to get in touch with your creative ancestors.

Visualisation meditation

There are ways to evoke the positive creative elements from your family tree. You can do this whether you know of a particularly inspirational family member or not. This meditation is to evoke the creative impulse within the family ancestral lineage.

Find photographs, heirlooms, family letters or information about the inspirational members in your family and place them on your altar. If you are adopted, choose a family member from either your biological family (if known) or adoptive family. If you do not know your ancestors and have not yet discovered any on your family tree that resonate with you, known or unknown, this exercise will enable you to conjure an ancestor who has your creative aspirations at heart.

Write down a list of your ambitions and goals.

THE PRACTICE

- Sit or lie down in silence. Breathe deeply once or twice, then bring your inspirational ancestor (known or unknown) into your mind. Ask that they be present for you and that they allow you to experience their creative power through your senses – vision, hearing, sensing and feeling.

- Visualise the energy of their creative power as coloured lights of bright yellow, green or blue, as these colours represent creative spiritual energy. Bring that light into your mind and body. Give thanks. Tell them you are here now, bearing witness to their power.

- Ask the ancestors to bless and inspire your goals. Ask any questions you have about your creativity. You can ask silently or out loud. Listen for the answers, which may come as fleeting thoughts, visions or feelings. When you are ready, thank them for their presence.

Reapply this meditation on a regular basis when you need help or inspiration or you have come to a creative block.

You can also learn to be an open channel for the ancestors' creativity. Light a candle, then sit down for a few minutes each day with your journal and simply allow words, symbols and ideas to flow on to the page without stopping to judge, censor or correct. If you don't know what to write, say so. Whatever comes to mind, write it down. Complete this task by giving thanks to your ever-present ancestral guardians. Remember you can call upon them: they will be there.

EIGHT
A Journey of Self-Discovery

Your ancestors have passed on to you much more than your physical attributes. The important facets of their lives, their successes, failures and temperaments are also reflected in you. You are bound to them through powerful psychic forces, even if you know nothing about them. Their experiences subtly shape your life.

DAVID FURLONG, AUTHOR OF *HEALING YOUR FAMILY PATTERNS*

Reading this book has probably brought back memories of your past and raised questions about how your ancestors may be affecting your life in the present. In this final chapter we suggest ways of looking at your own ancestral inheritance and embarking on a journey of self-discovery that will connect you with your ancestral roots. The journey begins with an exploration of your family's history. The second exercise will help you discover your own history from the time of your conception, and the third examines your cultural heritage.

Whenever we look back at our ancestral history we can usually recognise some of the many ways in which our ancestors still have an influence on us today. This knowledge may help us to admit what we like and dislike about ourselves and our family and then forgive and embrace some of the more negative aspects. Healing and transformation are more likely to happen if we can see what we have inherited and accept that sometimes our behaviour is a reflection of our more negative ancestors. In the final exercise, we acknowledge the positive aspects of the physical, psychological and emotional predilections that we have inherited and look at ways of unlocking our spiritual and creative potential through working on some of the negative effects of our inheritance.

Family history: who are your ancestors?

Our knowledge of our family's history is often limited to the very recent past and the memories of those who are still living. Few of us are lucky enough ever to have been in close communion with great-grandparents, and we are likely to see much less of our grandparents than people did in the past, when they lived close to one another all their lives.

If we live near to our relations, we can see all our own potential for craziness and foibles by looking at the mad aunt or eccentric grandfather, but it is far more difficult to see who we are, who we take after, what the predilections are in our family when we don't have regular contact with other family members. As you discover your ancestors' stories, you begin to see physical similarities and notice recurring patterns of behaviour. And you may well discover ancestors with whom you feel some special connection.

Finding out about our ancestors' lives and the struggles they faced will often help us understand why we are the way we are. In due course you may want to make an exhaustive search of your family tree, but looking closely at the history of the most recent two or three generations is sufficient as a starting point: you will soon discover family patterns and recognise inherited personality traits.

If you think you already know quite a lot about your immediate family, check whether what you know is actually true. I first started looking into the life of my grandfather after my cousin showed me the letters he had written to my grandmother from prison. It turned out that what I was told as a child (or thought I had been told) was not quite accurate. When you unearth the truth about an ancestor, rather than the myths, you may well discover surprising and enlightening characters. My experience has been that the more I found out about my grandfather's life the more fascinating it became and the more I felt emotionally connected with him.

The first steps are simply to find out and record as much as you can about your immediate family, starting with yourself, your parents and their parents (your grandparents). Making an ancestral notebook is a useful way to begin. Gather the documents and memorabilia you already have and get together with other branches of the family to pool information and share memories. If you want to go deeper and further in your research, check whether anyone else in other branches of your family has done it before. Are they still working on it? To find out, register with the various genealogical societies and then explore the many websites that have records accessible to the public.

Everyone must follow the same basic steps in genealogy, whatever nation or ethnic group they come from, but there are specific websites for different groups. Always be aware of the spelling of your surname and remember that most surnames will have variants. Keep an open mind on spelling, not only with people's names but with place names too.

If you are adopted, you can choose whether to research your adopted family or the family of your birth. If you have a strong spiritual link to your adoptive family, then treat that family as your true ancestral lineage. On the other hand, if you always felt like an outsider with your adoptive family there may be value in finding your birth parents and getting to know as much as possible about them.

Ancestral notebook

Buy a notebook to accompany you on your journey and use it to record all the information you find about your family and its history. Writing everything down in one place will provide the raw data for drawing up a family tree, if you decide to do this, and you will also be able to see familial patterns emerging from the mists of time.

Begin with yourself and the facts as you know them: dates of birth, baptism, marriage and where these events happened; details of the houses you lived in and the schools you attended. Write your own short autobiography, including any significant incidents involving members of your family, but also record any thoughts and feelings that come to mind about other people, such as schoolfriends or neighbours. This will be an interesting record for you in later life, as well as for future generations, and it may also help you identify events and relationships that you would like to explore further.

Repeat the process for your parents and your grandparents. If you don't already have documents to work from, check the accuracy of the various dates and events online if you can. Work back from one generation to the next, starting on a new page for each family member with a copy of a photograph, and including details of when and where people died, if relevant.

As well as noting the important dates, write down anything you find out about each person's physicality, personality, characteristics, profession and status, and tell their story as far as you know it. This information may come from personal knowledge or from other people's stories and descriptions, as well as from photographs, hobby collections and heirlooms. Ask friends and family what they remember about deceased members of the family: what were they like when they were alive, and what were their best and worst traits? Once you have recorded everything about your immediate family, you may want to start looking back at earlier generations.

An ancestral notebook is also a good place to record any dreams, insights and inspiration you have while exploring your ancestral lineage and connecting with past members of your family. These thoughts and ideas will be useful when working the later exercises in this chapter.

Family patterns

Members of our family may tell us that we take after someone in a previous generation. Perhaps we have the same colour hair, distinctive nose or unusually large hands. These are obvious examples of a physical characteristic that has been directly inherited through our

DNA. But, as you will have discovered when reading the stories in this book, there are plenty of other patterns to look for.

These inherited patterns may begin with a traumatic or significant event in someone's life. Did any of your ancestors die young from illness, in war or occupational hazard? Did they make advantageous marriages or have ruinous divorces? Is there a history of financial losses or successful entrepreneurship? Inheritance of wealth can have a powerful, often detrimental, effect on descendants and there are endless stories of great riches being accumulated by one generation, only to be lost within a couple more generations.

Families often have easily recognisable patterns of physical and mental health. Explore these systematically, beginning with your parents and going as far back as you can. Death certificates will include information about how your ancestors died; look for any repeating patterns and inherited conditions such as heart and blood disorders, heritable cancers or degenerative diseases. Do you know if any of your ancestors suffered from mental health problems or were admitted to an asylum? Conditions to look out for include suicide and depression.

You may be able to work out what kind of life your ancestors led from clues in the family tree. A traumatic childhood event such as being orphaned at a young age can leave someone with difficulties being demonstrative or loving and this will be reflected in the way they respond emotionally when they have a family of their own. If an ancestor ended life as a pauper, the next generation may have an expectation of failure and hardship. The experience of each generation affects those that follow, both directly and indirectly, so looking back into your family tree may well help answer some of the questions you have about your family's emotional make-up and your own inherited traits.

Many families have a puzzle in their past. There may have been whispers from elderly relatives about someone airbrushed from the family's official history, or an issue they refuse to discuss. Perhaps you are not aware of any shame or scandal but your research reveals incidents that were swept under the ancestral carpet: secret marriages, bigamy, suicide, illegitimacy, the loss of a baby, criminal behaviour, even murder. Discovering family secrets is fascinating but with it comes the responsibility of deciding what to reveal to other members of your family: it can be an emotional process to learn that your maternal great-grandmother lived in poverty in the tenements of Glasgow and died from tuberculosis, or that your great-great-grandfather was a slave owner. Researching relevant bits of history can put past family events in context and give you a sense of how they might have affected your ancestors at the time – and how they can still be felt in your family's lives today.

Laying out your family tree

As you accumulate more and more information about your family, it can be helpful to buy a big piece of white paper and begin to work on a physical representation of your family tree. This exercise will give you a clearer picture of the paths your family has taken in its journey leading to you: seeing where you have come from may help you understand where you are now.

Begin at the bottom of the card with a photograph of yourself, together with pictures of your brothers and sisters, if any. Place photographs of your mother and your father above these, and of their parents above them, and so on.

Write down what you have been able to find out about each

person next to their picture: where they were born, how they died and where, their education, their physical characteristics, any illnesses or other physical challenges, any positive and negative traits, their profession, careers or hobbies, religious beliefs and financial status. Note any historical events or incidents that affected them during their lifetime, as well as patterns that are repeated generation on generation.

Look at all the information you have managed to collate as if you are seeing it for the first time. What impressions hit you as you lay out your family tree? How do these people make you feel when looking back at you? What do you sense coming from their faces? Who else is in the picture with them and what is their relationship? Doing this will help you identify the relatives with whom you have a particularly close spiritual connection and who could perhaps become your guardian ancestors (see page 58). These may be people you already knew or ancestors from the distant past. With their help we can often find ways to heal family trauma and inherited behaviour and belief systems: we can help ourselves heal and also prevent the same issues being passed down to our descendants.

Write down your thoughts and impressions in your ancestral notebook, as you may need to refer to them in the later exercises which look at embracing the good in your family inheritance and healing what is not so good.

Personal history: who are you?

As we grow older and start to make choices about how to live our lives, perhaps discovering new talents and passions or reaching dead ends and battling with insecurity or addictions, we may start to yearn to find out who we really are.

In many ways, we are less free than we think. Our personalities are formed and moulded by our family history. This begins from the moment we are conceived, leading through our early childhood into adulthood. By understanding what we have inherited, we can begin to regain our freedom to choose our behaviour.

Mandala exercises

A mandala is a traditional spiritual tool, used for meditation and ritual, and is normally set out as three concentric circles to represent different physical or spiritual dimensions. In these exercises we are looking at your personal, family and ancestral issues.

The exercises begin with drawing the outlines of two mandalas (as shown in figs. 1 and 2), using a large piece of paper or card for each one. Do not include any of the writing shown on the mandalas: this is for your guidance when doing the exercises.

To complete the first mandala, find photographs of yourself as a baby, in early childhood, adolescence and adulthood, as well as photographs of your parents when you were either just born or in early childhood and at the present time, siblings, grandparents, partners and so on. Position the photographs in the boxes inside the mandala, as shown.

Spend a few minutes looking thoughtfully at the photographs. What do you see about yourself in these pictures and what do you feel about your parents and other members of your family? Think about the people who influenced you when you were growing up. Then use the questions in the box overleaf to help you think about yourself and your relationships with your family, writing down how you feel and what you think in your ancestral notebook. Try to be honest and truthful, spontaneous and intuitive in your responses.

WHO ARE YOU?

Laying out your photographs

Fig. 1

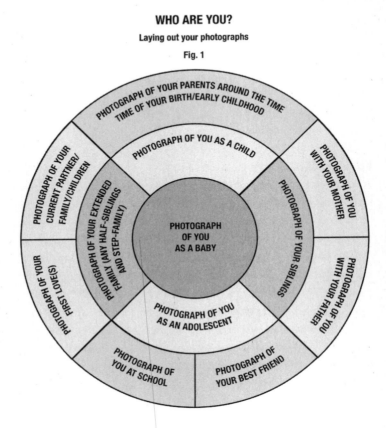

You may need a few sittings before you are able to answer all the questions. As a way of jogging your memory, you may find it helpful to look back at earlier chapters which describe other people's childhood experiences. See what comes up as you think about your past.

Reflecting on these questions about your beginnings will bring you into contact with your earliest memories and may shed new light

WHO ARE YOU?

Fig. 2

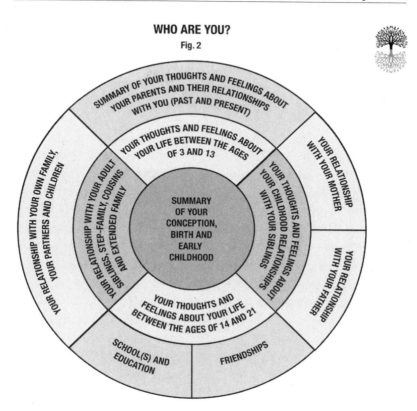

on your parental relationships. For example, I know I was deeply loved as a child but I later found out that I was illegitimate and my existence had forced my parents into marriage. When I discovered this, I began to understand the reason for the complex relationship that I had with my mother, who must have been emotionally distraught throughout pregnancy and birth. I realized that part of her subconsciously blamed me for what was a long and tumultuous relationship with my father. Consequently, it has taken me a long time to find my own healthy emotional foundations.

You may have had a happy childhood and a wonderful relationship with your parents and subsequently with your own family. But this is not always the case. Be honest with yourself when you look back into your childhood, even if you think you cannot remember much. Although we may not be aware of it, we are all affected by the emotions surrounding us at our conception and birth. We are marked by the relationship we witness in our parents and by our experiences with our siblings. If you have been adopted, you may have limited information about your birth family, but you can still use the questions as a tool for exploring your personal history.

When you have answered all the questions, put the mandala with the photographs next to the second mandala. Write a summary of your thoughts for each section on your second mandala, then meditate on your responses and how you have been affected. At the end of the chapter we will look at some ways to heal any issues there may be.

To complete this part of the journey, place the two mandalas on your ancestral altar. Light a candle and say a prayer to your ancestors for guidance and protection in your ancestral healing process, for yourself and for your family. Ask the ancestors for acceptance, forgiveness, healing, inspiration and transformation. Every evening for the next seven days, light a candle and repeat your prayer. Spend a few moments of contemplation in front of the mandalas, and open your heart and mind for change. After a week, remove the mandalas and keep them in a safe place with your ancestral notebook, to refer back to when you wish to work through new issues in your relationships with your family.

CONCEPTION

- What do you know about your conception? What have you been told about the emotions and status of your parents at the time of your conception? For example, were you a long-awaited miracle or a 'happy accident'? Or perhaps you were illegitimate and the subject of some scandal in your family?

- Were there any miscarriages, abortions, stillborn births or premature deaths of siblings before you were conceived?

BIRTH

- What was happening in your family at the time of your birth? How old were your parents? Were they happy or stressed? Did they feel overwhelmed by responsibility or were they pleased and proud?

- Was your birth easy or difficult?

- Are you the firstborn, a middle child or the baby of the family? Or are you an only child?

- What generation were you born into? (A baby boomer born after the Second World War or one of the Thatcher generation, for example.)

CHILDHOOD

- Who was the main provider, nurturer, caretaker in your early childhood?

- What were your relationships like with other siblings, half-siblings or step-siblings? Which of them did you get on with best, and why? What did you have in common?

- If you are an only child, who did you play with during your childhood?

- How did your parents get on with one another? And with other members of their generation?

- How do different members of your family remember your childhood? (Their memories may be different from yours.)

- Look back into your childhood between the ages of 3 and 13 and then between 14 and 21 – what was happening in your family life at these times, and were there any differences between early childhood and adolescence?

PARENTS

- Describe the relationship with your mother as you were growing up: was it warm or distant, encouraging or competitive, relaxed or controlling?

- Has it changed over the years? How old were you when it was at its most difficult?

- What do you know about your mother's childhood? In what ways do you resemble your mother? What have you inherited which reminds you of her?

- Look at your relationship with your father in the same way. How were you fathered as a child? Was your father present in the home or away a lot? Was he able to show his emotions?

- If you can, ask your father about his relationship with his own father – did he suffer any trauma such as war or sudden financial loss that may have affected his lineage?

- Do you feel you have to live up to your parents' expectations? Are you constantly trying to be the 'perfect' daughter/son, or do your parents love you as you are? Do you feel competitive with them or at ease?

- What do you most admire in your parents? What do you find most annoying? Do you recognise yourself in your parents?

FRIENDS

- How did you get on with people at primary school and then at secondary school? Did your home life affect your relationships with friends at school?

- Did you prefer family or friends when you were growing up? And now?

PARTNERS AND CHILDREN

- How has your relationship with your parents affected your relationships with partners? Have you consciously or unconsciously chosen people who are like your mother/father or unlike them? Is your relationship with your partner similar to your parents' relationship?

- If you have children, has your relationship with your partner changed since they were born?

- Have your own childhood experiences affected the way you are bringing up your children? Do you see your relationship with your parents reflected in your behaviour with your children?

Cultural history: what are your family's attitudes and beliefs?

Our values, beliefs and attitudes are part of the cultural heritage that is passed down to us through the family, and this exercise is an opportunity to identify your own family's rules and values. We are constantly affected by this history, although the effects may be unconscious or subliminal. Acknowledging these influences can help us gain a deeper understanding of ourselves. There will be some aspects that can help us grow and others that no longer support our creative and spiritual life. Once we know which parts of our ancient family codes are not helpful to us, we can begin to heal, accept or manage them.

Work through the questions below, answering them as truthfully as you can and recording the answers in your journal. Almost all aspects of your life will have been affected by attitudes picked up in the family home and, as with the previous exercise, these questions are not exhaustive, but are intended as a starting point for recognising your family's values.

Concentrate on those elements that are particularly important in your family and explore the attitudes and beliefs surrounding them. For example, your family might have a perfectly balanced attitude to money but an emotional response to food and weight. Then look through the notes you have made about your family tree and see if you can find the historical origins of these belief systems. List relevant names, dates and events; observe how these issues changed your family's lives and how some of these rules have become your family's belief systems today.

When you begin to see the rules and values that are operating within your family, you can decide which ones you are happy to keep and which you would prefer to change: you can start to lessen the dysfunction of some of the more negative family systems and restore a better balance in your own family.

Money Is your family parsimonious or profligate? Did you have a sense of abundance when you were growing up or did your family always seem to be short of cash?

Food Were meal times for gathering or gobbling? Did you eat in front of the television or sitting round a table? Did table manners matter? Were you forced to eat everything on your plate or were you allowed to choose what you ate? Was your mother constantly on a diet and counting calories?

Intimacy Are you a physically intimate family? Were there lots of hugs when you were growing up or more distance? Were your parents open or secretive about sex?

Discipline Were your parents strict or did they let you get on with things? Has discipline hindered or supported your life choices?

Gender Are sons and daughters treated differently in your family? Is there equality among you all or are there different expectations about education, marriage and inheritance?

Ambition Is there a sense that 'our family always gets ahead' or is naked ambition considered crass? Are your family supportive of natural talent or are they discouraging?

Work Do people in your family tend to be workaholics or lazy bones? Does your family have a strong work ethic or are they more relaxed?

Religion How does your family's religious background affect your human nature and belief systems? If you have children, do you teach them cultural beliefs? Or maintain any traditional or sacred practices at home?

Marriage Do marriages tend to be long and strong or is divorce and separation the norm? Have historical ancestors been illegitimate or had bigamous marriages? Do people in your family tend to marry young or wait? Is marriage something to which you all aspire or to be avoided at all costs? In divorce, did your parents manage to keep the children together in large extended families or was it acrimonious?

Making changes

It is time to look back at the work you have done in this chapter on the physical, emotional and cultural influences and memories that may be affecting you and take some simple steps to begin the healing process. You may feel happy to continue this process on your own, but this could also be a good moment to think about seeking advice and support from a professional therapist who specialises in family issues (see page 254).

Deciding which aspects of your inheritance you would like to change will take time. It is important to be honest and realistic about what you can change and what you cannot, and you need to know what you truly want for yourself and your descendants. One way to become clear about this is to spend an hour with your journal each week, looking back at everything you have discovered about your family heritage and making lists of the positive and negative attributes that you have inherited. This will help you become more aware of the elements you are happy to keep and those that you would prefer to lose.

Your physical appearance and constitution are simple starting points. You may well be happy with the colour of your eyes and hair, the shape of your hands and face, and your physical stamina and strength. Or you may wish some features of your inheritance had been different.

Now go back through the emotional and psychological issues that you have uncovered when looking at your family of origin and again write down a list of the positive and negative traits you have inherited. For instance, in my family we tend to be stubborn and self-righteous but also very generous, kind and supportive. Other positive attributes might include courage, intelligence, grace and compassion, while negative ones could be arrogance, jealousy, greed or aggression.

Finally, what would you like to heal or change about your family's attitudes and beliefs? List three positive aspects of your family's values and three negative ones.

Once you have made your lists of positives and negatives, look at them again and decide whether it is possible to change the negatives or whether you may have to accept them or learn to manage them. And if there are negatives that you can change, you then have to consider how willing you are to do whatever is necessary to make those changes happen. You can use the mandala on page 255 to help you summarise the healing that you would like to take place.

Before you start work on making changes, conduct a ritual to celebrate your rebirth. This could be a simple act of going to your favourite place in the landscape, writing a letter to your ancestors, telling them what you are healing and what other steps you are taking, burying the letter in the earth and placing a plant or tree above it as a gift of rebirth for yourself and for your ancestors.

Then you need to create some baby steps in your daily life to help manage the process of empowering the positive by taking the best that you have inherited from your family and releasing and letting go of the negative influences that no longer serve you.

- Make a list of the changes that you are happy to work on over the next twelve months and set yourself some practical goals.

- Visualise the transformations that can take place for yourself and for your family.

- Work on some of the exercises that we have suggested throughout the book. You can do the cutting ties or forgiveness meditation exercises (see pages 173 and 124) to help alleviate the inherited shadows that burden you.

- Pray and meditate about yourself and your family. Reflect on the aspects that you love and those you do not like and work on forgiving and accepting them. You can work through the positive aspects by conducting the visualisation meditation (see page 230).

- Connect with your ancestors every day (see page 55). Go on pilgrimages to see how your ancestors lived. Visit their graves and honour their memory.

- Make regular entries in your journal every day or week about the changes you are seeing in yourself and in your immediate family.

PILGRIMAGE

If you know where any of your ancestors came from, plan your own pilgrimage and invite other members of your family to share the journey of exploration. This is a way of reaching down the bloodlines, often more than one generation. You may not know exactly where your parents or grandparents came from, but a general area will do, as the atmosphere and the history of the place will pull you back to the memories of your own ancestral family.

When you arrive, honour your family's spiritual connection with the land: visit a local church or place of worship, a sacred grove or well. If you are lucky enough to find out where members of your family are buried, visit the site and conduct a simple ceremony. Say a prayer to your ancient ancestors, either out loud or in quiet contemplation: a prayer for their peace, protection and guidance. Light a candle, burn some incense and leave a gift that feels appropriate: money, food, water, something from your home, flowers or a plant. Honour and remember your family's heritage.

After twelve months, repeat the rebirth ritual, and then look back and check how this journey has been for you. You can also use the mandalas as a reference point and refer back to the notes you made in your ancestral journal.

Go to your sacred space, centre yourself with a few deep breaths, open your journal and answer the simple questions below to measure how far you have come since you first read this book. Trust your instincts when you answer. Follow whatever thought first comes to mind and notice how your body responds to the questions – are you irritated, bored and angry or happy and joyful?

- How do I feel in this moment?

- Am I now being true to myself?

- Is there something else I want to be doing with my life?

- What do I yearn for right now?

- What do I still need to heal for my family to flourish and grow?

- What do I need to do to be fully alive right now?

- Am I comfortable with my body? My mind? My emotions?

- What can I change?

- What are my goals for the next twelve months?

Give thanks to the ancestors for the progress you have made since beginning this process of healing your family inheritance.

HEALING YOUR ANCESTRAL INHERITANCE
Fig. 3

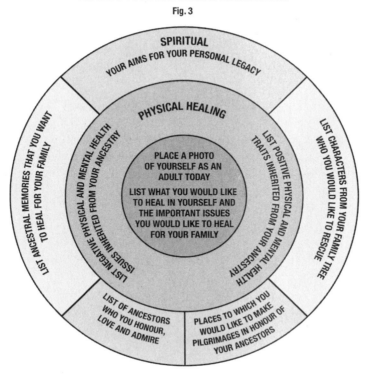

SPIRITUAL
YOUR AIMS FOR YOUR PERSONAL LEGACY

PHYSICAL HEALING

PLACE A PHOTO OF YOURSELF AS AN ADULT TODAY

LIST WHAT YOU WOULD LIKE TO HEAL IN YOURSELF AND THE IMPORTANT ISSUES YOU WOULD LIKE TO HEAL FOR YOUR FAMILY

LIST ANCESTRAL MEMORIES THAT YOU WANT TO HEAL FOR YOUR FAMILY

LIST NEGATIVE PHYSICAL AND MENTAL HEALTH ISSUES INHERITED FROM YOUR ANCESTRY

LIST POSITIVE PHYSICAL AND MENTAL HEALTH TRAITS INHERITED FROM YOUR ANCESTRY

LIST CHARACTERS FROM YOUR FAMILY TREE WHO YOU WOULD LIKE TO RESCUE

LIST OF ANCESTORS WHO YOU HONOUR, LOVE AND ADMIRE

PLACES TO WHICH YOU WOULD LIKE TO MAKE PILGRIMAGES IN HONOUR OF YOUR ANCESTORS

Epilogue: Legacy

We come to earth as blood family to follow our ancestors' steps who yearn for us to evolve and live out our legacy for the children of tomorrow and for our mother earth as the continuum goes on.

CARLA ESTEVEZ

Now we have rediscovered the stories of our ancestors, we can consider the story we will leave for our descendants. What would we like it to be? How would we like to be remembered? What gifts will we leave? What is our legacy? As ancestors-in-waiting we are already creating the story of the future.

Our legacy stretches from the material to the spiritual. The distribution of the wealth we have made in our lifetime is one way of helping our descendants: the family home, some heirloom pieces of jewellery, stocks and shares and a trust fund are our material legacy. Distributed fairly and with due care these things can be immensely helpful to those we leave behind. But there are so many other ways that we leave a legacy.

First, we bequeath to our descendants their own ancestral continuum when we remember, honour and forgive our ancestors. The understanding of who they are and where they come from gives our descendants the firm foundation on which to build their lives.

We also leave a legacy in the way we live our own lives and achieve our potential. As we begin to see ourselves as a link in the chain of evolution in our families, we can see the ways that we have inherited the memories of our parents and grandparents, their rules, values and traditions, and how we love both unconditionally and with conditions from our ancestral lineage. Then we can see clearly what we wish to pass on and what we do not.

Therapist Agatha Rodgers describes brilliantly the journey of personal liberation we take when we are able to gather the best of our ancestors' legacy and leave the rest. 'For years I struggled with the sense of rejection that one side of my ancestors projected on to the other: the battle between the free spirit and the landlord, the land farers and the land holders. Now I acknowledge the gifts, the shadow and light of both. Having detached from the social conditioning that they both enforce, I am using those as ingredients to transform the opposites into elements of a more creative and wholesome expression of myself.'

And by discovering the stories of our ancestors, we realise that our own story is also an important part of our legacy. As we look back at the lives of our family, we can see they have all left a little of their lives inside us; the laughter and tears, love and affection, kindness, respect, truth, honour and peace, as well as anger, turmoil and suffering. Their lives can give us that little bit of courage, strength and determination to lead us on our own journey with a new perspective. It is important to value our struggles and successes and still respect and be proud of who we are, however messy our lives might seem, as we too will pass on all our wisdom and life experiences – the culmination of our struggles and achievements – to inspire those who follow us.

Our personal journey will have long-range effects on our children's lives and all of our descendants. Just as if we have thrown a pebble into a still pond, the ripples expand from us into the lives of others who are a part of our family tree and outward into the ancestral continuum of humankind. We are leaving the footprints for those who will come after us, making their own mistakes and courageous decisions in a different time, creating a different history.

How will you be remembered? This is, already, a powerful legacy.

Your story

A friend discovered her father's diary after he died. He had kept it for more than thirty years of his life. It is a treasured gift for the whole family, far more valuable than any financial or material inheritance.

Your story is important. Write your story down as you consider how you want to be remembered. When you finish reading this book, write a biography of your own life with photographs, memories, personal and cherished items that you can put together in a box for your descendants. This can be passed on to each generation as a legacy of remembrance and honour.

You could also write a letter to your children (real or imagined) as though you were writing to your younger self, describing the lessons that you have learnt in your life. Include a list of the events in your life that have been significant and transformed you: rites of passage, historical and monumental moments, places that you have travelled to, people you have loved and lost, your education, career, small events that have touched you and times when you have touched others.

Imagine how they will remember you as a son, daughter, brother, sister, wife, husband, mother, father and grandparent. What do you want your legacy to be, how do you want to inspire you descendants?

- What are the things that you wish to be remembered for?

- How does your family see you and the way you have led your life?

- Who in your family do you wish to heal your relationship with and what are the steps you can take to do this?

- What would you like to heal in yourself in connection with what you have inherited from your family?

- What would you like to change before it is too late?

- What would you still like to achieve?

From our ancestors to our descendants

The perception of foundation is of great importance in our time, for we are coming to the end of a world cycle, and everything is beginning to unravel. We cannot begin the next cycle of ages without a foundation on which to build. And that foundation will begin to take form with the reconsideration of the wisdom held by the ancestors.

HALE MAKUA, HAWAIIAN WISDOM KEEPER

As we look back into our family tree we will all find an extraordinary ancestor whose life touches us so deeply that we can personally take their legacy into our own. If it sounds rather too reverential to discover and then be proud to be the descendant of a great ancestor, it is because my grandfather who was six feet tall with piercing eyes was a heroic character in every sense. Not only was he a judge, known for his fairness and honesty, he was also a fervent Socialist who campaigned for free education for everyone. As he looks down at his family, with the birth of another generation, I wonder how proud he must be that his short life was not wasted, seeing how many of his great-grandchildren have branched out to lead lives as lawyers, scientists and engineers, taking his legacy further down the family tree.

Katy found her great-grandmother's diaries of her time as a nurse during the First World War and discovered how courageous and compassionate she was, and how much of her character she has inherited. 'It's immensely inspiring to have such a brave woman in your family. It makes you feel like you can do anything. I'm sure that I'm so strong and independent because of her.' Now she is naming her own daughter Constance after her, passing on her great-grandmother's legacy to the next generation.

And singer Eliza Carthy has named her second daughter Florence, in honour of her extraordinary great-grandmother who took in her three grandchildren when their parents died young. Living alone in the austerity of post-war Britain, Florence managed to provide for them as well as her surviving children by setting up her own business. 'I gave my daughter her name to give her the

strength of character, warmth and love that I know my great-grandmother had.'

We receive gifts from ancestors we have never known and also
from those whom we knew when they were alive. As we were com-
pleting this book my Aunt Teresa, daughter of my grandfather,
passed away. Just as her father, my grandfather, had inspired me, so
too did she. She was an epitome of feminism – political, radical and
rebellious, a woman of her time. She taught me never to fear any
challenges, to be independent and to know that I had the capacity to
achieve my dreams. She gave me courage, strength and truth. These
extraordinary ancestors become the spiritual guardians who take care
of us, protect and guide us and our children. I am now leaving
behind my own legacy for my family, by my actions, my words and
my loving presence within the family.

So how do we follow an ancestor who changed the course of history or was a reflection of their generation?

- How do we take on the legacy of the lives of our many, many
  ancestors who we have never known?
- How do we remember and honour them and how does that
  inspire us to be who we are?
- As we look back into our family, who is the ancestor or deceased
  family member who you have chosen as your inspiration?

Our descendants

Every child born wants to know who they are, or where
they come from and where they are going. The rich-
ness of knowing who your ancestors are gives you the
road to walk on.

GRANDMOTHER PAULINE, MAORI ELDER

We bequeath to our descendants a material, emotional, cultural and spiritual legacy. Dying a good death is a part of that legacy. Of course, we cannot plan the way in which we die, but we can prepare for its eventuality in many ways and the more prepared we are, the better it is for our descendants. How do you want to die? If you want to die at home surrounded by family with soft candlelight and Mozart, then you need to prepare for it. Create a living will that tells your descendants how much medical intervention you do or don't want. Do you want to be buried or cremated? What music do you want at your funeral? All these details help your loved ones deal with your death in the best possible way. If you have assets, make sure that you have distributed them fairly and get professional help if you need to.

Making a will is not just a legal exercise that begins and ends with people's needs and desires. It is an emotional journey that can be an opportunity to sit together as a family and discuss the issues involved, air past jealousies and misunderstandings face to face, listen to each other and make a plan.

Before Terry's father died he discussed his plans because he wanted his passing to be as peaceful and easy for his family as possible. On the other hand a friend's mother never wanted to talk about what would happen to her property and personal items, so she died without a will and her three daughters fell out immediately after her death and are still unable to reconcile their differences. Unfortunately, some 70 per cent of us die without a will and those who do make one rarely discuss it before they die, causing years of misunderstanding and suffering for their descendants. Sit down and discuss how you would like to leave your personal items and property, and say what you would like at your funeral. Your descendants will thank you for it.

And while we are contemplating our death, it is a good moment to consider all the other gifts that we give to our descendants, so that they may enjoy their lives and enhance the lives of their children and so on down the family line. Think about yourself as a parent. What are the traditions and rituals that you are continuing and passing on to your children? These are a precious part of the rich inheritance that you carry and bequeath to your descendants.

Family traditions become the touchstones of our lives and those of our descendants: fathers teaching sons to play ball and taking them to their first football match; grandmother helping us bake our first cake; family rituals like playing Scrabble, going on walks, making bonfires or just piling together on the sofa watching TV on a Saturday night. These precious memories of family time remind your children of their childhood which they will recreate for their children and so on.

And when we continue the cultural and spiritual rituals that we love about our family, we also continue the best of who we are: stockings at Christmas; feasts at Thanksgiving, Norooz, Shabbat and the Sunday roast; splitting the wishbone for good luck; planting daffodil bulbs in winter and saying prayers at bedtime.

It is in our families that we learn how to be human as we pass on our family's values and virtues: how to respect money and each other, form intimate relationships and become parents ourselves.

In relationship to the continuum of our ancestry, we carry the light of all our ancestors and, whether or not we have children ourselves, we give this light to our descendants. The light of our ancestors shows up in so many ways: our creative gifts and talents; our determination and courage; our humour and lightness of heart. As we look at our descendants we can see how each child carries a different gift or talent and how we may nurture them so they too can share their gifts with the world.

- What do you want your descendants to inherit from you?
- What gifts and talents have you passed on to your descendants?
- What physical and psychological traits have they inherited from the family?
- What have your children or descendants inherited from your parents, grandparents and great-grandparents?

Our universal legacy

> Humankind has not woven the web of life. We are but one thread within it. Whatever we do to the web, we do to ourselves. All things are bound together. All things are interconnected.
>
> CHIEF SEATTLE (1854)

As we connect to our own family tree we also connect to the vast family tree of humanity. The genome has proven we are more alike than we are different and the internet has brought us closer than our ancestors would have thought possible. Now we have an opportunity to take collective responsibility for our planet. For without the earth we have no legacy.

Indigenous peoples around the world talk of prophesies laid down hundreds of years ago by our ancestors about a time when humanity would make a collective decision to care about our planet and each other. In all the prophecies, humanity makes a quantum shift in consciousness, a massive evolutionary leap where we learn to take our place as custodians of the earth.

Grandmother Pauline Tangiora, an elder of Scottish and Maori descent and activist for indigenous rights, reminds us that we are all indigenous when she says, 'Through your umbilical cord you are connected to generations of your ancestors going right back to the beginning. Europeans have as much knowledge as indigenous people. You know your ancestors' stories and their ways. The knowledge of how to take care of the earth has never been lost, it was just sleeping.'

When we experience tough times – world economic collapse, war, dislocation and natural disasters – we can feel as though our foundations are shaken to the core. But crisis is also an opportunity for change and humans are uniquely able to adapt to a changing world, and to change the world to render it liveable. So as we reach another crossroads in our evolution, we can turn to the ancestral wisdom in our DNA and our cultural memory.

When the environmental crisis feels overwhelming, it is good to remember we are descended from people who depended on the earth for sustenance and health, from hunters and gatherers, from farmers and fisherman, from people who knew how to live in balance with the elements and maintain sustainability. We carry as our birthright a love of the natural world that we can call on to protect our legacy and ensure our children and children's children inherit the beauty and abundance of the earth.

And when our world leaders squander this inheritance, we can remember our ancestors who changed the world and do our bit to protect it. No action is too small. Every choice makes a difference, from our buying habits to donating money to environmental groups.

Our ancestors challenged the status quo at every turn and brought us civil rights, votes for women, rights for children. History has proven again and again that our collective desire for freedom and dignity is greater than the forces that seek to oppress us. But usually it is crises that spur us into action as a collective, cohesive whole. And that is when we are most effective.

Above all we need to realise that we are all custodians of the earth, caretakers of a legacy that began on the First Day. Our oldest ancestors awoke to sunshine and slept under the stars shining bright in a dark sky, drank from clear streams and freely hunted game, honoured the forests and the mountains, the rivers and the seas and saw the hand of the Creator in all things.

The concept that our actions today will be felt by the next seven generations – originated by the Great Law of the Iroquois Confederacy – is one that encourages us to look ahead beyond our own personal self-interest and towards protecting the earth for generations to come. And when we see ourselves as a link in this chain of evolution we can begin to see that our lives have the potential to bring change and transformation across time. Stop for a moment and consider this question: If we were dreaming the world that our descendants will inherit, what would it look like? How would we plant the seeds to create it? What kind of world do we want to bequeath to them?

The ancestral journey – we are the descendants

The gift the ancestor guardians gave to us was the knowledge of our purpose and our destiny and this shared wisdom is two-fold: First we were brought here to enjoy ourselves – to grow, increase and become more than we were in the beauty of nature on this wonderful world. And second, we are to remember our divine origins through the experience of love for one another. This is it. This is what we are here to experience. All the rest, all our work and accomplishments, our successes and failures, our families and friends, everything we do and become in our lives is simply the river of experience that carries us, the background against which we struggle or with which we flow as we learn our life lessons and transform into our once and future selves.

HALE MAKUA

What you can see you can heal,
what you deny will haunt you,
what you have the courage to change will transform you,
what you bury will challenge you,
what you unbury will become gifts and inherited talents.
Give thanks to your ancestors,
give thanks to your spirit and
give thanks to your descendants.

Recommended reading

Traditional wisdom

Laura Amazzone, *Goddess Durga and Sacred Female Power* (Hamilton Books, 2010)

Joseph Campbell, *The Power of Myth* (Anchor Books, 1991)

Chartwell Dutiro and Keith Howard (eds), *Zimbabwean Mbira Music on an International Stage: Chartwell Dutiro's Life in Music* (Ashgate, 2007)

Brooke Medicine Eagle, *Buffalo Woman Comes Singing* (Ballantine Books, 1991)

Mircea Eliade, *Shamanism* (Routledge, 1989)

Clarissa Pinkola Estés, *Women Who Run With the Wolves* (Random House, 1992)

Michael Ortiz Hill and Mandaza Augustine Kandemwa, *Twin from Another Tribe: The Story of Two Shamanic Healers from Africa and North America* (Quest Books, 2007)

Robert Lawlor, *Voices of the First Day* (Inner Traditions, 1992)

Denise Linn, *Descendants: tracking the past, healing the future* (Rider Books, 1998)

——, *Altars* (Rider, 1999)

Credo Mutwa, *Indaba, My Children: African Tribal History, Legends, Customs and Religious Beliefs* (Payback Press, 1998)

——, *Song of the Stars* (Barrytown, 1996)

John G. Neihardt, *Black Elk Speaks* (Washington Square Press, 1932)

Jamie Sams, *The 13 Original Clan Mothers* (HarperCollins, 1993)

Malidoma Somé, *The Healing Wisdom of Africa* (Thorsons, 1998)

——, *Of Water and the Spirit* (Tarcher/Putnam, 1998)

——, *Ritual: Power, Healing and Community* (Penguin, 1997)

Dhyani Ywahoo, *Voices of our Ancestors* (Shambhala, 1987)

Personal histories and self-help

Isabel Allende, *Paula* (HarperCollins, 1995)

——, *The Sum of Our Days* (Harper Perennial, 2008)

Edward Bell, *Slaves in the Family* (Ballantine Books, 1998)

Liza Campbell, *Title Deeds* (Doubleday, 2006)

Sara Connell, *Bringing in Finn: An Extraordinary Surrogacy Story* (Seal Press, 2012)

Karen Doherty and Georgia Coleridge, *Seven Secrets of Successful Parenting* (Bantam Press, 2008)

Niravi B. Payne and Brenda Lane Richardson, *The Fertility Solution* (Thorsons, 2002)

M. Scott Peck, *In Search of Stones* (Hyperion, 1995)

Max du Preez, *Pale Native* (Zebra Press, 2003)

Anne Ancelin Schutzenberger, *The Ancestor Syndrome:*

Transgenerational Psychotherapy and the Hidden Links in the Family Tree (Routledge, 1998)
Noel Tovey, *Little Black Bastard* (Hodder, 2004)
Cami Walker, *29 Gifts: How a Month of Giving Can Change your Life* (Da Capo Press, 2009)
Sjanie Hugo Wurlitzer, *The Fertile Body Method* (Crown House Publishing, 2009)

The aftermath of war

Marianne Hirsch, *Family Frames: Photography, Narrative and Postmemory* (Harvard University Press, 1997)
Leila Levinson, *Gated Grief: The Daughter of a GI Liberator Faces Her Inheritance of Trauma* (Cable Publishing, 2011)
Micheline Aharonian Marcom, *Three Apples Fell From Heaven* (novel) (Riverhead Books, 2001)
Paul Preston, *The Concise History of the Civil War* (Fontana Press Spanish, 1996)
Dr Edward Tick, *War and the Soul* (Quest Books, 2005)
Giles Tremlett, *Ghosts of Spain* (Faber & Faber, 2008)

Death and dying

P.M.H. Atwater, *Beyond the Light* (HarperCollins, 1994)
E.A. Wallis Budge, *Osiris: The Egyptian Religion of Resurrection* (Kessinger, 2003)
Stanislav and Christina Grof, *Beyond Death* (Thames and Hudson, 1980)
David Kessler, *The Rights of the Dying* (Vermillion, 1997)

Soygal Rinpoche, *The Tibetan Book of Living and Dying* (Harper Collins, 1992)

Rodney Smith, *Lessons from the Dying* (Wisdom Publications, 1998)

Robert Thurman, *Tibetan Book of the Dead* (Thorsons, 1998)

Felicity Warner, *Gentle Dying* (Hay House, 2008)

——, *Safe Journey Home* (Hay House, 2011)

Researching family history

Mark Herber, *Ancestral Trails: The Complete Guide to British Genealogy and Family History* (Sutton, 2004)

David Hey (ed.), *The Oxford Companion to Family and Local History* (Oxford University Press, 2010)

Roger Kershaw, *Migration Records: A Guide for Family Historians* (The National Archives, 2009)

Megan Smolenyak, *Who Do You Think You Are?: The Essential Guide to Tracing your Family History* (Viking, 2009)

Alan Stewart, *Grow Your Own Family Tree: The Easy Guide to Researching Family History* (Penguin, 2008)

Bryan Sykes, *The Seven Daughters of Eve* (W.W. Norton & Company, 2002)

For information about genealogy websites please go to www.theancestralcontinuum.com

Acknowledgements

Natalia O'Sullivan

An immense gratitude to my husband, Terry, for his dedication to our work and his infinite support and loving encouragement, which has enabled us to complete this book with wisdom, grace and compassion that merits its content and intentions.

Thank you to my children, Sequoia, Ossian and Bede, and my mother Purita for all their patience and loving support. To my brother Paul, for his great sense of humour and helping me to keep it real, and to his wife Natalie and my nephews Marcus and Louis. To all my cousins and extended family – the Silva Pandos in Spain and to the Kovacs family in Hungary.

And in honour and remembrance to:

my grandmother Purificación (Pura) Durán García
her daughters Patucas Pando Silva and Teresa Pando Keenan

my father Ferenc Josef Kovacs
my grandparents Josef and Rosalia Kovacs

Nicola Graydon Harris

To my loving husband, Rob Harris, for his great heart, his emotional support and his beautiful way with words. To my grandparents, William Graydon and Johanna van Eeden, who gave me Africa and Ireland; to Willard Charles Williams for his encouragement of me; and to the indomitable Gertrude Rachel Corrie and my beautiful great-Aunt Lully.

I am so thankful to my family. To my mother Mary, my sister Sheena and my brother Steve for all the love, loss and learning on our journey together. To my brother-in-law Marco, for always being a champion of my writing. To my nephews Ben, Caspar and Felix and my stepsons Sam and Casey who are a constant inspiration and fill me with hope for the future. And to my aunt Patricia and all our cousins in South Africa.

Further acknowledgements

To Kerri Sharp, who answered the call of her ancestors in commissioning this book and whose unswerving commitment and dedication kept us on the path to completion.

To Sally Partington and Serena Dilnot, our genius editors, who patiently guided us towards the true vision of this book.

To Emma Maloney for her wisdom and compassionate mediation that helped us to remain focused.

To John Ridout and Lizzie Myers who, at a critical point in the book, let us write in their panelled room at Hunstile Farm.

A special thanks to our friends and contributors who gave their time, advice and generosity on this project: Isabel Allende, Andura, Laura Amazzone, Alexandra Asseily, Laureen Bishop, Sara Bran, Anita Bains, Mbali Creazzo, Georgia Coleridge, Janine Clements, Sara Connell, Wendy Cazzolato, Chartwell Dutiro, Dr Wendy Denning, Zena Duez, Gaye Donaldson, Carla Esteves, June Elleni-Laine, Angel Gutierrez and Mona Gutierrez, Katherine Hooker, James Hyman, Laura Hayward, Taryn Jacobs, Mandaza Augustine Kandemwa, Leila Levinson, Heidi Lawson, Myrna Clarice Munchus, Pia Mellody, Lady Agatha Rodgers, Lei'ohu Ryder, Malidoma Somé, David Sye, Donna Stewart, Pauline Tangiora, Maestro Tlakaelel, Dr Edward Tick, Jack Van Der Linde, Evelyne Valabregue, Sjanie, Hugo Wurlitzer, Cami Walker, Dr Felicity Warner and Angela Watkins.

And a big thank you to all our friends, clients and students from around the world for their unconditional support and their ancestors' stories, many of which are being told here for the first time. Those stories are the heart of this book, and even the ones we have been unable to include or acknowledge are a part of its life force. We are eternally grateful: Rose Anson, Tania Attar, Arit Anderson, Kate Brown, Dawn Burnett, Mercedes and Peggy Betancor, Holly Barren, Eliza Carthy, Sarah Coles, Gustavo Pernas Cora, Tim Carver, Margaret Donald, Max du Preez, Acacio Da Silva, Abigail and Sue Dooley, Harmoni Everett, Kathy and Amy Eldon, Laura Facey, Zac Goldsmith, Charlotte Gray, Marianne Goldstein, Danny Hansford,

Romana Hoossein, Stephanie Hutchinson, Lucia Hargasova, Katherine Hooker, Andrea Hook, David and Dian James, Philip Jones, Kalpana Jogia, Lorraine Kinman, Ann Kelson, Johnny Lyne Pirkis, Tai and Vivien Long, Mark and Wendy Lee, Charlie Lipton, Gary Morecombe, José Molinos, Max Milligan, Myrna Clarice Munchus, Briony Newman, Nef'fahtiti, Ouhi Uluhogian, Nader and Pegah Pakfar, Ross and Louisa Pepperell, Nick Powell, Kalinka Poullain-Jacobs, Genna Preston, Steve Rodgers, Rupesh Srivastava, Julian and Carina Reed, Aggie Richards, Drew Spencer, Greta Scacchi, Bernard Valentine Slack, Baz Stamboulin, Behiye Suren, Margarita Teijeiro, Katie Tinne, Philip Turland, Noel Tovey, Mary Vanderhook, Camilla Walker, Kate Watson and Emma Westcott.

For information about The Ancestral Continuum workshops and seminars and to contact Natalia O'Sullivan or Nicola Graydon, email them at theancestralcontinuum.com.

To find out more about our contributors and the therapies they offer for ancestral healing and indigenous wisdom, go to www.theancestralcontinuum.com, where you will find personal interviews and website contact details.

Sources

Every reasonable effort has been made to trace copyright holders of material reproduced in this book, but if any have been inadvertently overlooked, the publishers will be glad to hear from them.

Introduction

Denise Linn, *Descendants: tracking the past, healing the future* (Rider Books, 1998)

One

Mandaza Augustine Kandemwa, interview with Nicola Graydon, 2011
James Shreve, 'The Greatest Journey', *National Geographic* (March 2006)
Rosanne Cash, interview with *The Courier* (uncredited) on Tour Scotland website at http://www.fife.50megs.com
Hugh Masakela, out-take from *Evening Standard* interview, Nicola Graydon (30 November 2000)

Two

Arseny Tarkovsky (tr. Virginia Rounding), excerpt from 'Life, Life' in *Life, Life: Selected Poems* (Crescent Moon, 2007)
Felicity Warner, interview with Natalia O'Sullivan
Credo Mutwa, *Song of the Stars* (Barrytown, 1995)

Three

Isabel Allende, interview with Nicola Graydon, 2011
Malidoma Somé, *The Healing Wisdom of Africa* (Thorsons, 1999)
Aubrey Burl, *Circles of Stone: The Prehistoric Rings of Britain and Ireland* (Weidenfeld & Nicolson, 1979)
Clarissa Pinkola Estés, *Women Who Run With the Wolves* (Random House, 1992)
Nancy Cobb, *In Lieu of Flowers* (Random House, 2002)
Gustavo Pernas Cora (tr. Manda Denton), 'The Good Memory'

Four

Anne Ancelin Schutzenberger, *The Ancestor Syndrome: Transgenerational Psychotherapy and the Hidden Links in the Family Tree* (Routledge, 1998)
Alexandra Asseily, 'Breaking the cycles of violence in Lebanon – and beyond' (Geurrand-Hermes Foundation for Peace Publishing, 2007)
Denise Linn, *Descendants: tracking the past, healing the future* (Rider Books, 1998)
Anne Ancelin Schutzenberger, op. cit.
Dr Christiane Northrup, foreword to *The Fertility Solution*, Niravi B. Payne and Brenda Lane Richardson (Thorsons, 2002)
Rosie Anson, interview with Natalia O'Sullivan
Dr Dan Booth Cohen, 'Are Constellations Alchemy?' at danbooth-cohen.blogspot.co.uk (2011)

Five

Marianne Hirsch, 'Surviving Images: Holocaust Photographs and the Work of Postmemory' (*The Yale Journal of Criticism* 14: 5–37)

Leila Levinson, *Gated Grief: The Daughter of a GI Concentration Camp Liberator Discovers a Legacy of Trauma* (Cable Publishing, 2011

Anthony Browne, interview in the *Guardian*, Sarah Crown (4 July 2009)

Shirley Williams, preface to *Testament of Youth*, Vera Brittain (Victor Gollancz, 1979)

Anne Ancelin Schutzenberger, *The Ancestor Syndrome* (Routledge, 1998)

Alcinda Honwana, 'Sealing the past, facing the future' (Accord Mozambique, 1998)

Dr Edward Tick, *War and the Soul: Healing our nation's veterans from post-traumatic stress disorder* (Quest Books, 2005) and interviews with Nicola Graydon

Bob Cagle, in 'Beyond PTSD: Soldiers Have Injured Souls', Diane Silver reporting for the Miller-McCune Report (1 September 2011)

Leila Levinson at www.veteranschildren.com

Hilde Schramm, interview in the *Guardian*, Henrik Hamren (18 April 2005)

Nancy Morejon, excerpt from 'Black Woman' in *Freedom in My Heart*, Cynthia Jacobs Carter (ed.) (National Geographic, 2009)

Dr Dan Booth Cohen, 'The Legacy of Slavery Workshop' at dan-boothcohen.blogspot.co.uk

Sid McNairy, interview with Nicola Graydon, 2011

White Feather, on Native Beliefs at http://home.earthlink.net/~tessia/Native.html

Max du Preez, *Pale Native* (Zebra Press, 2003) summarised by Nicola Graydon with permission

Noel Tovey, interview with Nicola Graydon, 2011

Kevin Rudd, http://www.theaustralian.com.au/news/nation/full-transcript-of-pms-speech/story-e6frg6nf-1111115543192

Six

Kitty Hagenbach, Parent Child Psychotherapist and Co-Founder of Babiesknow, written for this book

Owen Sheers, from 'Not Yet My Mother', *The Blue Book* (Seren, 2000)

Richard Llewellyn, *How Green Is My Valley* (Penguin Classics, 2001)

Jonathan Haidt, *The Happiness Hypothesis* (Random House, 2006)

Greta Scacchi, out-takes from interview in *The Times*, Body and Soul, Nicola Graydon (2006)

Seven

Sir Arthur Conan Doyle, 'The Adventure of the Empty House' in *The Return of Sherlock Holmes* (1903)

Shirley Abbott, *The Bookmaker's Daughter: A Memory Unbound* (Ticknor & Fields, 1991)

Dr Glayde Whitney, *The Evil Gene* by Frank Stephenson (Published in Research in Review, 1996)

Pema Chodron at Pemachodronfoundation.org

David Sye, interview with Nicola Graydon

Joseph Campbell, *The Power of Myth* (Anchor Books, 1991)

Emilio and Marco Nella, 'The blade runners', interview in the *Independent*, Clare Longrigg (February 2005)

Epilogue

Carla Estevez, written for this book

Hale Makua in *The Bowl of Light: Ancestral Wisdom from a Hawaiian Shaman,* Dr Hank Wesselman (Sounds True, 2011)

Grandmother Pauline, interview with Nicola Graydon

Chief Seattle, from his famous speech of 1854

Hale Makua, op.cit.

About the Authors

Natalia O'Sullivan has worked for the past twenty years as a holistic therapist, healer, psychic and spiritual counsellor who combines modern psychological thinking with ancient wisdom. She has studied psychology and mastered various holistic arts such as shiatsu, massage, reflexology, crystal healing and shamanic traditions.

She runs busy healing practices in London, Somerset, Spain, Ireland and Los Angeles. With her husband, Terry, she is the co-founder and creator of The Soul Rescuers Foundation Course, a professional certification training programme which draws on ancient spiritual traditions with a contemporary approach. Trained Soul Rescue practitioners specialize in sacred and ancestral healing, spirit release and land healing. To find out about the course and to contact qualified practitioners of Soul Rescue and Ancestral Healing go to www.soulrescuers.com

With her husband, Natalia is the co-creator of The Sacred Healer Retreats based in south west England. Sacred Healer Retreats' approach to healing and therapy is based on Celtic Shamanism, which encourages a spiritual connection with nature to restore our link with our ancestors. These retreats cultivate a sacred understanding of the earth and honour and communicate with the ancestors. The retreats are based on shamanic practices such as vision quest, rituals, meditation, ancestral healing and communication. For retreat programmes and information visit www.soulrescuers.com or contact Natalia@sacredhealers.co.uk

As a writer, she is the co-author (with Terry O'Sullivan) of the seminal spiritual manual *Soul Rescuers* (Thorsons, 1999) and author of *Do It Yourself Psychic Power* (Element, 2002) and *Mind Power* (Ryland, Peters & Small, 2003). She has regularly written articles on spiritual and alternative health issues for *Spirit*, *Light* and *Living the Field* magazines.

Nicola Graydon is a writer, journalist and occasional broadcaster. Her first-ever story for the *Sunday Times* Magazine on the San Bushmen of South Africa ignited her passion for exploring indigenous wisdom traditions. She has also written for the Saturday *Telegraph* magazine, *The Times*, *Daily Mail*, *Mail on Sunday*, *Evening Standard*, *Eve*, *Marie Claire*, *Harpers Bazaar* and the *Ecologist*. After marrying film publicist Rob Harris and decamping to Los Angeles, her journalism took her to several film sets including *Troy*, *Syriana* and *Blood Diamond*. She has interviewed actors George Clooney, Morgan Freeman, Sharon Stone, Jennifer Connolly, Thandie Newton, Tony Curtis and Emma Thompson, and been shark diving with Leonardo di Caprio. Currently living between London and Los Angeles, she continues to explore the use of ancient spiritual technologies in the modern world.

www.nicolagraydon.com